AF769754

UNDERWATER EAR AND NOSE CARE

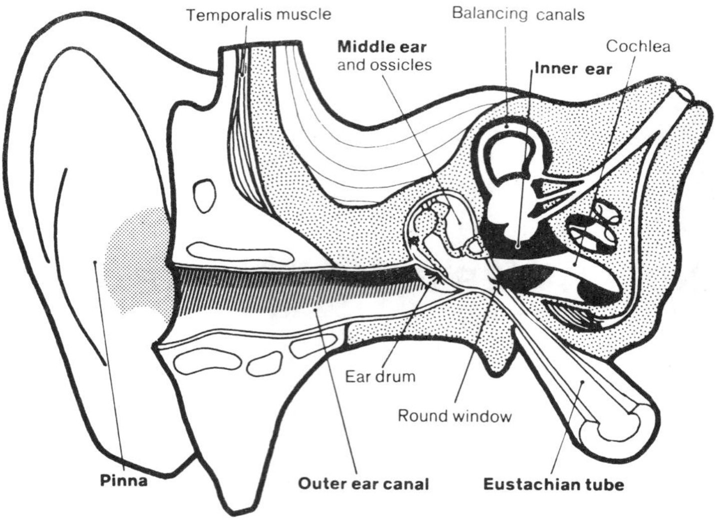

The three parts of the ear -
Outer, Middle, Inner.

Underwater Ear and Nose Care

By

Noel Roydhouse V.R.D.

Ch.M. (U.Otago), F.R.C.S. (Eng.), F.R.A.C.S.

Hon. Surgeon-Captain, R.N.Z.N.V.R.
R.N.Z.N. Hyperbaric and Diving Medicine Unit.
Editor, New Zealand Journal of Sports Medicine.
Member, South Pacific Underwater Medicine Society.
National Medical Panel, New Zealand Surf Life-Saving Association.
Honorary Medical Officer, Auckland Surf Life-Saving Association and Waitemata Amateur Swimming and Surf Life-Saving Club.
Member, Veteran's Underwater Club.

Dedicated to the Te Ngaere Bay Social Club

Foreword by Surgeon-Commodore A.G. Slark R.N.Z.N. (Rtd).

Cover design by Bob and Matthew Allen, Idea Design, Parnell, Auckland

First published 1981

Second Edition 1993

Published by

BEST PUBLISHING COMPANY

2355 North Steves Boulevard
PO Box 30100, Flagstaff, Arizona, 86003-0100 U.S.A.

Telephone 602/527-1055 FAX 602/526-0370

also for: Medical Books (N.Z.) Ltd.,
P.O. Box 8565 Auckland,
New Zealand

92-80522
Library of Congress Catalog Card
Number
ISBN 0-941332-23-3
International Standard Book Number

Printed in New Zealand by The Phoenix Printing Co. Ltd., Auckland

Foreword

It is with much pleasure that I am able to introduce a new edition of this valuable book on the care of the diver's ears and nose. First published over two decades ago as "Scuba diving and the ear nose and throat," it has throughout this time been of great help to divers themselves, diving instructors, and medical practitioners. It assists them to avoid the difficulties that may effect this particularly sensitive system of the body, and to manage the problems when they do occur. The very first thing that a novice diver must do is learn how to adjust to the changing pressures associated with going underwater. Being able to equalize pressure each side of ear drum is a fundamental prerequisite before any diving can proceed. The explanations given here together with the diagrams will continue to be most helpful to those of us who are involved with the teaching and advising of divers.

The inner ear is a critical target area for serious decompression illness as well as direct pressure injuries and these are well described and illustrated by case histories.

Noel Roydhouse continues to provide both amateur and professional divers with the benefits of his long experience as an active diver and a researcher. The field of barotrauma is little understood in his speciality but he is able to describe it to us in this book with precision and clarity.

A.G. SLARK M.B., B.S., M.F.O.M., M.R.C.G.P., D.P.H., D.I.H., D.Obst.
Surgeon-Commodore, RNZN (Rtd)

Preface

The years since the book on "Scuba Diving and the Ear, Nose and Throat" was first published (1971) have seen a great increase in the number of underwater enthusiasts. Accordingly the number of divers who have developed problems as a result of venturing into the pressurized environment of what can be unfriendly depths has also increased.

While the text has been prepared specifically for those that venture recreationally and professionally underwater, it is hoped that others working in this field such as medical practitioners, physical educationists and especially instructors will find it useful if they study it in detail. The former would then understand more about the problems that their diving patients present to them, and the lattermost would be able to give better and more factual information to their student divers. Too many instructors make up their own versions of anatomy and physiology. I was taken aback when a recent intern asked me why do swimmers have a complete coating of wax in their outer ear canal. This incorrect information had been learnt from his instructor and he accepted it completely and uncritically despite six years of scientific medical training.

At the same time I have been able to collect together many more cases of divers' problems, not only those using tanks, the SCUBA divers, but also those whose troubles arose from snorkeling or swimming. In fact, the chapter on the "Outer Ear" was based on an original investigation into surface swimmers in 1969. Combining all three groups has given me the experience of diagnosing and treating about 2,000 cases including over 1,200 Scuba divers, for their water sport-induced problems. During this period, investigations have been made into various particular aspects of ear and nose problems and the main investigations have been included in their own chapters. The complete list of publications in the scientific medical and other media is appended. As some of these extra chapters are an elaboration of points already made in the text, to avoid distorting the simplistic approach, they have been given their own chapters. This has lead to some repetition but increased elaboration does enhance the educative value. It is the care of the ears and nose which is important to underwater swimmers and it is hope that many will find that attention to the described prevention measures will enable them to get more enjoyment out of their exciting recreation.

The case histories, usually simplified, are included in the text to illustrate major points. Because they are actual experiences they show how often the story of an incident can be complex and not as simple as the diver may think. Also I must apologise for the use of the male nomenclature in the text but I will point out that my male diving patients out number the females by about five to one.

The revision of the 1981 printing began in the Fiji Islands whilst I was holidaying at the scuba diving resort of Matagi Island adjacent to Taveuni Island and its world class diving. My thanks go to the proprietors Noel and Flora Douglas, and their staff for the good food that kept me going and the lovely relaxing surroundings of the Matagi Island Resort.

Noel Roydhouse
November 1993

Contents

Page ii		The Three Parts of the ear.
Page v		Foreword
Page vii		Preface
Page 2	Chapter 1	THE UNDERWATER DIVER. Fitness for Diving.
Page 5	Chapter 2	THE EUSTACHIAN TUBE. Situation, Visualisation, Inflation, Function, Relief of factors hindering opening.
Page 14	Chapter 3	THE MIDDLE EAR. Anatomy, Abscess in the ear, Myringotomy, Middle ear catarrh, Clearing the ears, Valsalva Manoeuvre, The Squeeze or Middle ear barotrauma.
Page 26	Chapter 4	THE OUTER EAR. Wax, Infection, Diffuse external otitis, Exostoses, Outer ear squeeze.
Page 36	Chapter 5	THE INNER EAR. Deafness, Dizziness, Alternobaric vertigo, Mastoid Cavities.
Page 49	Chapter 6	THE NOSE AND SINUSES. The Common Cold, Allergic Rhinitis, "Sinus", Nose Bleeds, Sinusitis, "Sinus Squeeze".
Page 60	Chapter 7	MOUTH AND THROAT. Lip Sores, Chronic Catarrh of the Throat, Adenoids.
Page 63	Chapter 8	THE JAWS AND HEADACHES. Headaches and Earache.
Page 70	Chapter 9	HOW TO CLEAR THE EARS or prevention of middle ear barotrauma.
Page 81	Chapter 10	VERTIGO, DIZZINESS AND NAUSEA.
Page 89	Chapter 11	THE JAW AND UNDERWATER SWIMMING.
Page 96	Chapter 12	THE SCUBA DIVING DOCTOR.
Page 108	Index	
Page 111	Glossary	
Page 112	Medicine list	
Page 114	Scuba diving scientific publications	

CHAPTER 1

THE UNDERWATER DIVER

Fitness for diving

The trained scuba diver is above average for emotional and psychological stability, and certainly does not suffer from claustrophobia, because of his acceptance of the underwater environment. He has a basic knowledge of the function and structure of the ear and nose, although sometimes he has added on his own misinterpretations, especially of the sinuses. He is an avid listener to details of this region to add to his imperfect knowledge. Because he spends a considerable amount of money and holiday time on his recreation, he is motivated to quick recovery so that he can continue diving. Indeed, he tends to continue diving during treatment, unless this is forcefully and strictly prohibited. His experience is that, although he is usually told to "stop diving" for his sports injury, he has kept on, and eventually recovered, so that his faith in his medical attendant can be undermined. He shares this mishandling of his sports injury with many of those injured in other sports and this need for understanding of sports injuries and the sportsman is one of the functions of the American College of Sports Medicine and the New Zealand Federation of Sports Medicine.

Although there are many more serious disorders arising as a result of diving other than ear, nose and sinus complaints, these latter three are more common. It is estimated that 80% of divers give up within 12 months of beginning diving. The common cause for a diver giving up would be the pain and discomfort caused by ear problems. Others give up because of the stress and their feeling of not being able to exercise self-control when things start to turn bad.

The Eustachian tube joins the back of the nose to the middle part of the ear. The underwater swimmer frequently has much trouble with these regions, a rare cause of trouble in a non-swimming adult. The function of the Eustachian tube is important and the decision whether a person is suitable for diving or not, either medically or by natural selection, relies on its tests of function being normal. Many of the adjustments to the respiratory systems of the aquatic mammals can be observed in the conditioned human but of all the adaptations, the human ear is that part which undergoes very little in the way of alteration to the underwater environment.

If when holding the nose pinched closed and with a swallow (Toynbee Manoeuvre) the prospective diver hears a click, pop or a feeling of pressure in his ears, then his Eustachian tube function is normal. Failure for this to happen does not indicate complete lack of function but auto-inflation (Valsalva Manoeuvre) with pressure should be tried (see Chapters 3 and 9). If the person cannot then clear his ears, if it is not due to faulty technique, then a specialist opinion should be obtained before declaring the person suitable for scuba diving. It is only rarely that the ear should be the cause of failure to pass a diving medical.

Because of this difficulty in equalizing pressure, the ear does suffer more often than other parts of the body, and although most of the damage is reversible, failure to receive the correct treatment may lead to permanent disability. If any deafness or noises in the ear develop urgent specialist treatment is required. Although the underwater environment has a very small safety margin, ear injuries rarely place the life of a diver in jeopardy.

Fitness for diving then comes into the orbit of the medical profession and a prospective diver should be referred for a medical examination as to whether he is a suitable person to take up underwater swimming. As far as the ears are concerned, there should be non-perforated eardrums in a normal outer ear canal (Fig. 4 p. 14). These drums should be freely mobile and of normal thickness. Thin eardrums which are floppy and easily ruptured, can be, but not always are, a contraindication to diving. There should be no active or chronic infection in or around the ears, and the hearing should be normal. If there is any defect then the opinion of an otologist or ear specialist, who has practical associations with diving, should be obtained.

For safety in diving there should be no great disorder of the nose, which is the case if the diver is a nosebreather. Some divers have badly blocked noses from allergy, but they have no difficulty with clearing the ears when diving. Accordingly a blocked nose is not necessarily a contraindication to a medical clearance for diving. If the diver is a mouthbreather, then a further opinion should be sought. If there is any history of purulent sinusitis or nose bleeding then a specialist should be seen. Most mouthbreathers have some degree of catarrh of the throat or chronic pharyngitis with an overactive gag reflex such that they gag and choke when a spatula is used to depress the tongue. With an irritable throat, a tendency to gagging, and likely similar irritability of the vocal cords, it could be hazardous, as undesirable coughing underwater can be caused by a cold re-expanded air of the scuba apparatus. This cold air is more a feature of the 2-stage reducing valves because of the final expansion occurring in the mouth-piece itself, with the cooling effect of the expanding air.

"Sinus" trouble is the folk medicine name for what was formerly known as nasal catarrh. However every person has their own definition of it which can vary from a cough to a dizzy turn. "Sinus" trouble is not written about in standard medical text books but is a folk medicine term commonly used by lay people. It consists of attacks of stuffiness of the nose with or without some excess mucus coming from the nose and some sneezing. Any associated face-ache or headache is likely to be due to jaw muscle tension. Although folk medicine "sinus" trouble may be allergic or stress related, the sinuses very rarely show any abnormality.

From the general point of view the heart and lungs should have no gross disorder. Certain conditions are an absolute contraindication to diving and this includes epilepsy, asthma, emphysema and a history of pneumothorax. A full account of fitness for diving is outside the scope of this volume.

In the end the major common problem to divers is to prevent middle ear squeeze by inflating the ear, also known as clearing the ears, performing Valsalva Manoeuvre or auto-inflating the ears. Toynbee Manoeuvre (p. 20) can usually be relied on for a test of normal Eustachian tube function. With experience any otologist or ear surgeon can make any diver's ear work safely underwater.

Key Points

1. Ear Nose and Sinus complaints are common in divers.
2. Eustachian tube is critical for easy and safe diving.
3. Beware of folk-medicine tales.
4. Any ear can be made fit for diving.

CHAPTER 2

THE EUSTACHIAN TUBE

Because the Eustachian tube plays the greatest part in all ear problems of diving it is dealt with in the first ear chapter. The tube was described by Bartolomeo Eustachio about 150 years before Valsalva worked on it. It is important to know something about it to understand the ear problems that happen.

Situation

The Eustachian tube average 1½ inches (4cms) in length, is situated in the middle of the head adjacent to the horizontal base of the skull (Fig. 1). The outer half, starting from the cavity of the middle ear, is in a canal in the bone of the base of the skull whilst the inner half is partly

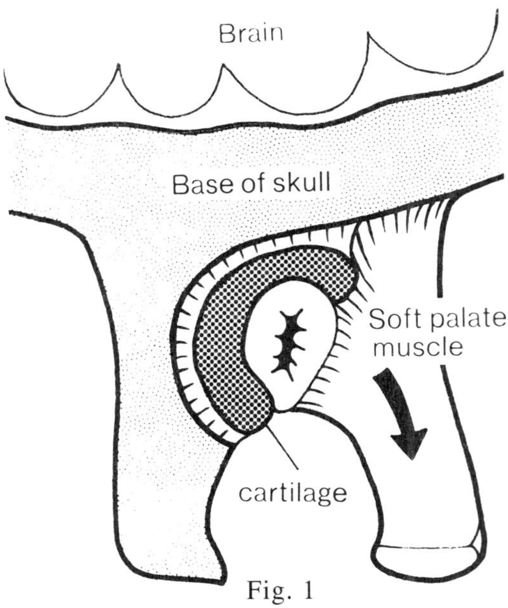

Fig. 1

Section through the cartilaginous part of the Eustachian tube. The soft palate muscle contracts on swallowing and pulls on the soft tissue front wall of the tube.

surrounded by cartilage and leads into the throat at the back of the nose, above the soft palate and thus normally out of sight. The cartilage is a C-shaped plate with the front open side being filled with soft fibrous tissues of the various muscles and the lining of the throat. The tube in the resting state is collapsed or closed, preventing the free passage of air both to or from the middle ear. Thus in the resting state the Eustachian tube is normally blocked. This blocked tube is physically blocked and this has no relationship with any blocked feeling felt in the ear. The muscle which lifts the soft palate upwards is attached to the soft tissue front of the Eustachian tube and when it contracts it pulls the tube open. Despite being placed deep in from the surface the Eustachian tube is accessible to the surgeon when required.

Approach and Visualisation

The inner end of the Eustachian tube projects up to 1/2 inch (1.2 cm) into that part of the throat at the back of the nose, known as the nasopharynx, where it can be seen with a mirror (Fig. 2) in the adult. The projecting part of the Eustachian tube is called the Eustachian cushion, and in the centre of which is its opening. Radiating from the Eustachian cushion can be seen the ridges of the muscles attached to it (Fig. 24 p. 76).

Inflation

The common methods of equalizing the pressure in the middle ear with the surrounding water pressure, are Valsalva Manoeuvre or auto-inflation, Frenzel Manoeuvre and the Twitch as described in Chapters 3 and 9. Politzerisation (Fig. 33 p. 100) is another clinical method used by doctors in their surgeries.

Function

The function of the Eustachian tube is to allow air into the middle ear, equalising the pressure of the middle ear with pressure at the back of the nose and thus the pressure of the surrounding environment - "clearing the ears". This is necessary under normal atmospheric conditions to allow the eardrum to vibrate freely to maintain normal hearing. Air is absorbed from the middle ear into the tissues around the middle ear cavity, lowering the air pressure. Replenishment occurs naturally with swallowing.

Disorders can develop when diving, because under these circumstances, pressure changes can be rapid and great in magnitude and the Eustachian tube must function readily. Submariners for example, practise compression to 600 feet in 30 seconds for escape procedures and must clear their ears continuously or they rupture their eardrums.

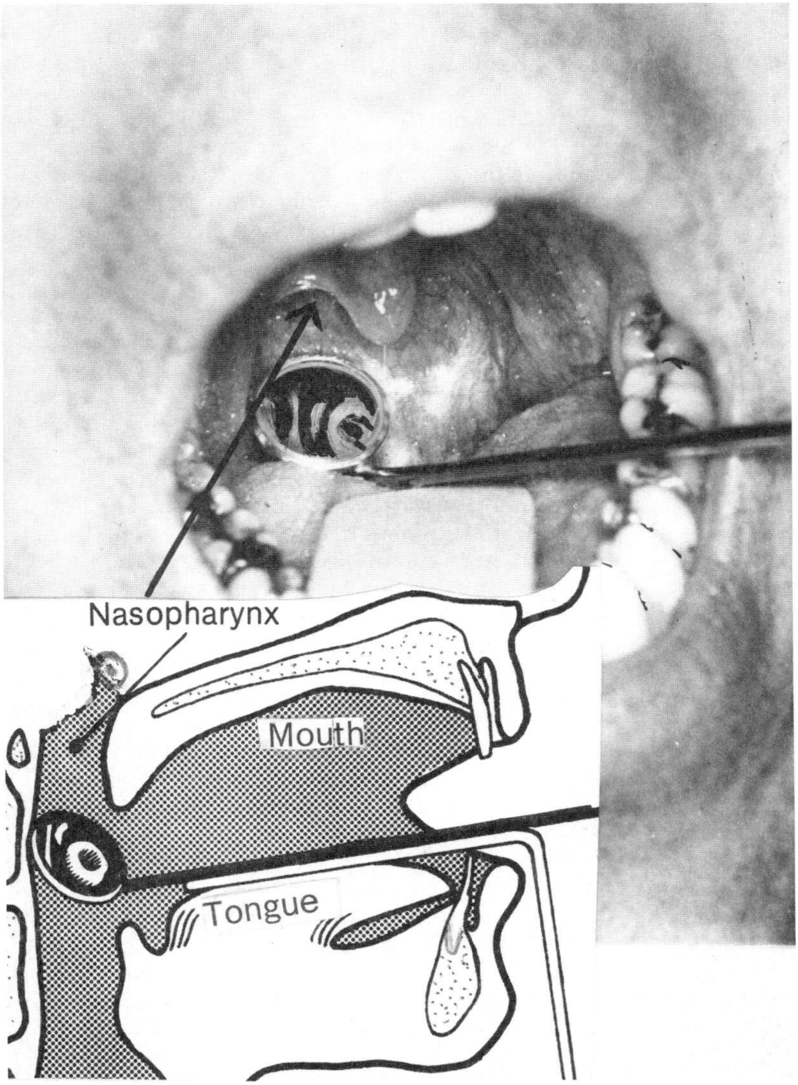

Fig. 2

Viewing the Eustachian tube and its opening, with a mirror in the back of the mouth. Adenoids can be seen if present.

How the Eustachian Tube Opens

The muscles moving the soft palate have the major part of their attachment to the front or soft-tissue side of the Eustachian tube. These muscles pass forwards and down the side wall of the upper throat or nasopharynx, turning inwards into the soft palate. When a person swallows the soft palate is pulled upwards by these muscles to close off the back of the nose. At the same time the soft tissue of the Eustachian tube is pulled forwards and down. As the cartilage of the tube is semi-rigid and attached to the base of the skull, this results in the opening of the tube to allow air to flow up or down as the air pressure dictates.

In the normal person without deafness, the Eustachian tube opens, on the average, with every second swallow. A normal person swallows, on the average, once every minute when awake, and once every five minutes when asleep. This rate of swallowing is fact, but is done mostly without the person being aware of it. So when diving, the major difficulty in "clearing the ears" or "equalising" is due to lack of correct technique as most people have a normally functioning Eustachian tube. To overcome this difficulty the diver should practise out of the water and clear his ears at least ten times daily. With such practice some divers quickly develop easy techniques, others still have trouble. With practising, people begin to become aware of the sensations in the back of the throat of the muscle movements and can then move onto the other methods described in Chapter 9.

How to Recognise that the Eustachian Tube has Opened

When the Eustachian tube opens there are sensations felt by the diver indicating that the tube has opened. These are felt even in a situation where there already exists equalisation of pressure between the back of the nose and the middle ear. These sensations are the same as with the Toynbee Manoeuvre or merely swallowing. If the opening of the Eustachian tube is accompanied by pressure equalisation then these sensations are usually greater. The sensation is usually described as a hiss, click, bang or pop or merely a feeling of pressure. In a diver who is feeling the pressure in his ears due to descent or is feeling a mild pain, the opening of the tube with pressure equalisation results in the loss of these sensations (Fig. 3).

Those whose ears clear with a hiss usually have some mild form of obstruction, usually due to congestion. The hiss is often heard with passage of air from the middle ear on ascent in such people. The click varies in intensity and it can be very faint. The sound can be mimicked by clicking two fingernails together or by moistening the thumb and index finger with saliva, holding them close to the ear, then separating them. This moist sound on separating the thumb and finger is similar to the soft click produced by the Eustachian tube opening.

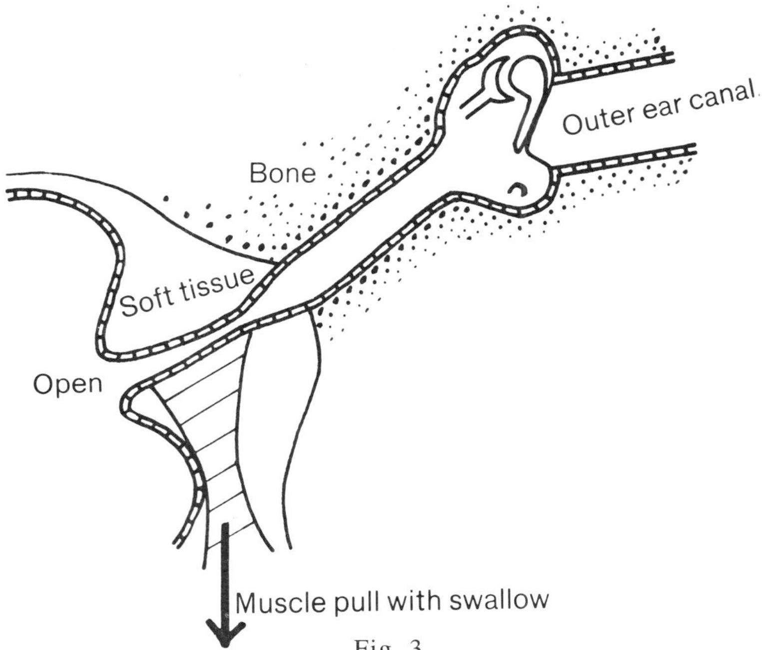

Fig. 3

The Eustachian tube is open because one of the swallowing muscles of the soft palate has pulled the soft tissue on the front of the tube. As it opens a "soft click" or other sound is heard by the diver.

Case History

Mr D.N. who had started a course for scuba divers was sent by his own doctor because he was having a variety of symptoms following his sessions in the pool and these were blamed on his inability to clear his ears on descent. He was asked to pinch his nose closed and to swallow. This is the Toynbee Manoeuvre for demonstrating normally functioning Eustachian tubes and described by Joseph Toynbee about 140 years ago. He was asked whether he heard a click or pop in his ears. He said he could hear his swallow but he could not say whether his ears cleared or not. With a stethoscope on his ears he listened to my ear whilst I opened my Eustachian tubes. He then admitted that that was the sound he heard and he didn't realise that such a small sound was all he was expected to hear when he cleared his ears. He then wet his thumb and index finger and listened to the sound as they separated. He agreed that this was what he heard when he performed the Toynbee Manoeuvre.

He rang up the next day to say that he had not had any trouble with his ears during the diving session but he had developed a pain in his neck, a throbbing headache in his temples, and when he arrived home he began vomiting. These and his previous anxieties were due to his mental tensions and apprehensions and he really did not seem psychologically suited for scuba diving.

Factors Hindering Eustachian Tube Opening

The factors which hinder the opening of the Eustachian tube are many. If the mucous membrane lining the tube is swollen or congested due to allergy, increase in the number of the mucous glands or due to inflammation such as the common cold, then there would be greater difficulty in opening what is, in the resting state, a non-open tube.

The nose is lined with a mucous membrane which produces mucus at the rate, in the average adult, of about 5 pints (3 litres) every 24 hours. Most of this is swallowed with the previously mentioned swallowing movements. Under the mucous membrane lining is a layer of tissue containing a plentiful supply of arteries, veins and large-type veins, normally empty. However they can be filled up with blood in a similar manner to the erectile tissue of the penis. When these large-type veins are filled, the lining of the nose takes up more room and there is less space for the air to pass through. At this stage the nose is said to be congested and it usually feels stuffy or blocked, as occurs in those with "sinus" trouble. This nasal congestion or filling of the blood vessels in the mucous membrane of the nose affects the lining of the Eustachian tube as well so that there is less room, normally less than 1/2mm, for air to pass up into the middle ear. Alcohol causes congestion of the mucous membrane of the nose and Eustachian tube, and such drink should be prohibited before diving. Congestion of the veins in the nose and the Eustachian tube occur when lying horizontal, such as when sunbathing, and this can lead to blockage of the tube.

The adenoids (Fig.20, p.62) and similar antibody-tissue in the Eustachian cushions, swell up with infections more so than other tissues. This antibody producing tissue around the Eustachian tube opening, which also may be widespread in the side and back walls of the nasopharynx, can press on the Eustachian tube from the outside. This prevents the tube from opening, preventing easy aeration or "clearing the ears".

Table 1
Factors Hindering Eustachian Tube Opening

Nasal catarrh (sinus trouble)	Infected rhinitis
Allergic rhinitis	Sinusitis
Alcohol	Lying flat
Adenoids	Adhesions
Deviated Nasal Septum	Vomero-ethmoid spur
Poor technique	Common cold

Following the inexpert removal of the adenoids in childhood, or following severe infections, adhesions or bands of scar tissue sometimes form between the Eustachian cushion and the roof of the throat or nasopharynx. These adhesions prevent easy working of the mechanism for opening the Eustachian tube.

Further hindrance to opening is caused by the pressure differential which builds up as a result of going underwater. The soft tissue of the inner half of the Eustachian tube, acts in a valvular manner and as the pressure rises it is compressed (Fig. 8). With the rise in the blocking pressure, greater effort is required to overcome the increasing compression. This pressure is equal to the column of water between the diver and the last level at which his ears were equalised. "Locking" or complete inability to open the tube can result if this pressure differential rises more than one quarter of an atmosphere (10 feet or 3 metres in water) and the diver should ascend until he can clear the ears easily. At the stage of "locking", the chest muscles do not have the strength to produce a high enough opening air pressure to counter the closing water pressure. More about Eustachian tube problems is given later on in Chapter 9.

Relief of Factors Hindering Eustachian Tube Opening

In order to counter these factors which reduce the ability to open the Eustachian tube, it is necessary to maintain the mucous membrane of the nose in a healthy state. Any nasal catarrh, commonly known as "sinus" trouble, whether due to allergic, stress or other reasons, should be treated. As far as the lymphoid or adenoid-type tissue is concerned, this always swells with infections and this swelling is one of the reasons for not diving with a cold. Any adhesions around the opening of the Eustachian tube can be broken under local anaesthetic. The index finger is gently introduced through the mouth and up behind the soft palate into the top of the throat or nasopharynx to break these.

Case History

Scuba diver D.M. reported to me because of difficulty in clearing his ears. Examination revealed no abnormality to account for the Eustachian tube not opening easily when required. However he allowed a look with a mirror into the back of the nose (nasopharynx) and there attached to his Eustachian cushions were adhesions or bands of scar tissue. His throat was anaesthetized and a finger inserted deep into his mouth. The finger was turned upwards and swept around the Eustachian cushion breaking the adhesions. A few weeks later he reported by telephone that he was now able to clear his ears with much less difficulty. Not "without difficulty" but he could descend much quicker without his previous ear pain.

As far as the pressure differential is concerned, it is necessary to "clear the ears" or put air into the middle ear at frequent intervals so as to prevent the pressure differential arising. In beginners and in those who have difficulty in clearing their ears, continual practice out of the water can improve the ability to inflate the middle ear. It is not unusual for the diver to inflate his ears, even before leaving the surface. This gives a higher pressure in the middle ear completely reversing any "squeeze" and giving time to get stabilised under the water before clearing is required again. Then the diver should be continually clearing his ears, every two feet of descent. Which of these factors are preventing the opening of the Eustachian tube, and which are not, is not always an easy matter to decide and usually several factors are in operation at any one time.

To make it easier to open the Eustachian tube, after carrying out the measures already outlined, the use of nasal decongestants is advised, taken twenty to thirty minutes before entering the water. The oral ones are tablets of pseudo-ephedrine (trade names: Sudomyl or Sudafed), or combinations such as Eskornade and Actifed. The modern antihistamines no longer have the undesirable side effects of causing drowsiness, and do clear allergic noses.

Because of the possible undesirable side-effects which vary from person to person, these medicines should be tried on a non-diving day, at the maximum dose. If there are no side effects then that decongestant is safe. In this way any reaction can be discovered before the diver is underwater, and another decongestant can be tried. Nasal sprays are effective and the metazolines, Otrivine, Drixine or Atomol, can be used before diving to clear nasal congestion.

Complications of Eustachian Tube Blockage

The Eustachian tube is the key to some of the permanent damage that divers do develop, and its function should be fully understood. Rupture of the eardrum can occur even after a ten foot (3m) dive if the tube will not open and allow inflation. Failure to clear the ears leads to the "squeeze" or barotrauma of the middle ear. Permanent deafness of the sensorineural type can occur in divers who have to use prolonged force in inflating their ears (see Chapter 5).

Case Histories

Mrs M.R. whose diving experience was recent, lost sight of her buddy when at about 80 feet (25m). She looked around and saw two divers about 10 feet below her, or so she thought. She descended to these divers who were actually at 120 feet (38m). Because she failed to inflate her buoyancy compensator she descended rapidly. To clear her ears she had to use prolonged and forceful blowing down the nose. On return to the boat she had minor loss of balance and the next day on the way back to Auckland she noted that her right ear was blocked and singing. Her own doctor said, "I told you so. Scuba diving is dangerous. You should not have taken it up. You have permanent deafness and nothing can be done for it". Her friend referred her to me.

She was operated on and her hearing returned but not quite to normal. Forceful inflation had ruptured the delicate Round Window membrane separating her middle and inner ears and the resultant outflow of fluid deranged the inner ear. At operation the membrane was repaired. She continued diving for at least another five years with no further problems according to a friend who came to see me.

Miss W.P. had completed her diving course despite having much difficulty in clearing her ears. Because of this difficulty when she did her sea test, she had descended down the anchor rope slowly so as to be able to clear her ears at leisure without having to worry about swimming. She was then advised by her instructor to seek help to improve her ability to clear her ears. Her technique was good but she had some stuffiness in her nose. After cautery under local anaesthetic, of the congested mucous membrane of the nose, under she was able to breathe through it more easily and clear her ears easily. She returned after four years for another unrelated condition and said how much easier clearing her ears had been right from the time of the minor operation. She also said that the procedure had also cured her "sinus" problem.

Case No. 41 came to see me for a check one year after his last dive which was to 40 feet (13 metres) at sea. He had experienced no pain on descent but on surfacing some blood came out of his left ear. He visited his family doctor who told him that he had a hole in his eardrum and this was probably true. He was also told that he was never to dive again. My examination revealed that the left eardrum appeared thicker and stronger than the right eardrum. He did however have a negative Toynbee Manoeuvre (p.20) due to some nasal congestion so this was treated successfully with cautery or heat treatment to the mucous membrane of his nose. He was then to go back diving and has not been seen since.

Key Points

1. The Eustachian tube normally opens with swallowing to allow air up into the middle ear.
2. There are factors which stop the Eustachian tube opening.
3. Failure to easily open the Eustachian tube leads to complications.

CHAPTER 3

THE MIDDLE EAR

The middle ear (Fig. 4) between the outer ear and the inner ear, is that part of the ear which is most commonly affected by disease. Many lay people mix up the middle ear with the inner ear but not divers. It is also most affected by the pressure changes during diving. To counter these pressure changes, the seal has large veins in the lining of the middle ear. On descent these veins fill with blood at pressure slightly greater than the external pressure. They gradually fill the air space in the middle ear maintaining automatic pressure equalisation.

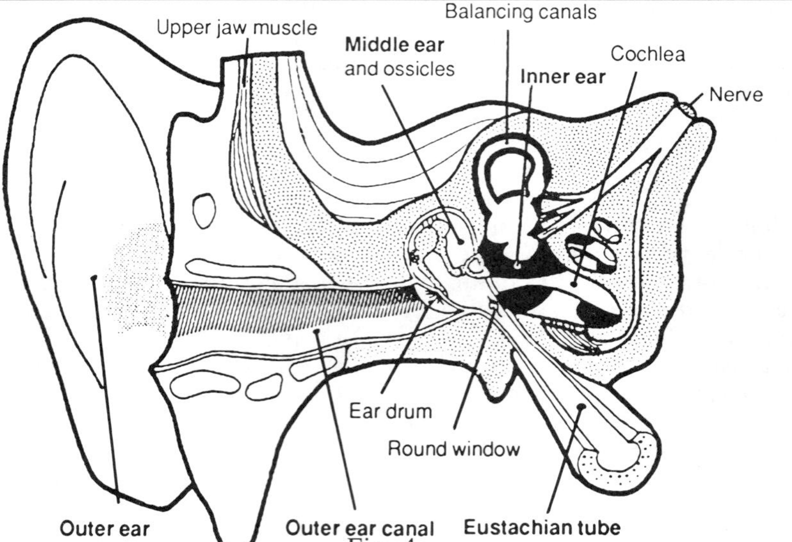

Fig. 4

Diagram of the three parts of the ear: Outer, Middle and Inner.

Because the middle ear transmits sound from the eardrum to the inner ear, by way of three little bones, any interference with its function and any infection in it, results in varying degrees of deafness. If your hearing is not affected and is normal then you do not have any significant disorder of the middle ear.

The important part of the middle ear is the eardrum, a circular vibrating structure situated at the inner end of the outer ear canal and whose inner end it completely seals off. It is thin, being of no greater thickness than tissue paper, but it is normally quite strong. The skin covering its outer surface is similar to and continuous with the skin of the outer ear canal. Its inner surface lining is continuous with the thin one cell thick lining of the middle ear. In between these two linings there is the important middle layer of fibrous or supporting tissue which gives strength to the eardrum. If this layer is missing, in part or whole, then the drum is seen to be very thin, it moves excessively and is liable to rupture or break with pressure changes. Repeated ear infections or long term mucous ear or "glue ear" in childhood causes this supporting tissue to gradually disappear. The eardrum is then thin and weak.

On the inner side of the eardrum is the middle ear, which contains the three little bones which transmit the vibrations of the eardrum to the Oval Window. From there the vibrations are carried into the inner ear where they are converted into the nerve impulses of hearing by the Hearing Cells. At the front of the middle ear chamber there is the opening of the Eustachian tube (Fig. 5), and leading out from the back of the middle ear at the top is another opening to the mastoid region. The mastoid region is above and behind the ear and is a honey-comb of cells filled with air. Rarely chronic infection develops in these air cells. Then mastoid operations are carried out to cure the infection.

Diseases of the Middle Ear

Infection of the middle ear, otitis media or an abscess in the ear, may be acute with pain or merely chronic with discharge of pus from the ear. The infection involves the whole of the middle ear cavity, thus affecting the eardrum and the little bones, with resultant deafness. It also can easily spread through into the mastoid region, and although this is a minor part in acute infection, it is more important in chronic infection.

ABSCESS IN THE EAR or acute purulent otitis media

This usually starts as a complication of another disorder such as a cold or sinusitis when the infection from the nose travels up the Eustachian tube. The spread of infection is aided by swimming and by incorrect nose blowing. With swimming the irritation of water in the nose aggravates the nasal infection and the pressure of the water can force infected mucus up the Eustachian tube. If there is a hole in the eardrum, water can wash infection through this into the middle ear. It is important for a person with a perforation to take precautions against ever getting water inside the ear. The result of the infection may just be marked congestion, or more commonly, pus is produced inside the middle ear. Such pus quickly gathers in the middle ear and if it cannot get away down the Eustachian tube, an abscess is then formed in the middle ear. If there is a perforation, the pus will come out of the outer ear canal as a creamy thick ear discharge often sticky in consistency. This abscess in the ear is more common in children but rare in adults.

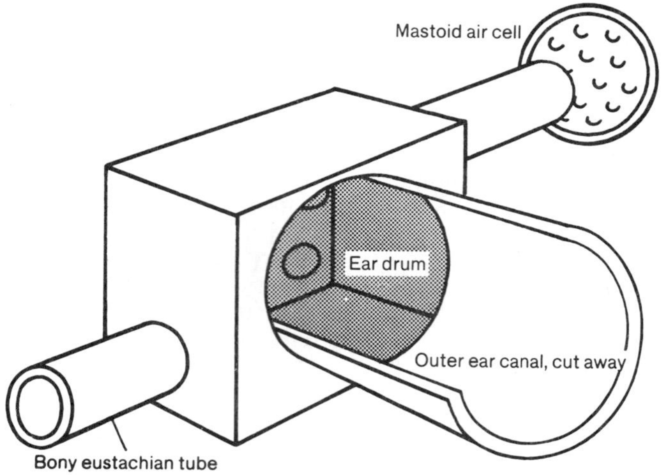

Fig. 5

Diagram of the middle ear and its connections. The eardrum seals off the outer ear canal from the middle ear. The only connection of the middle ear to the outside is through the Eustachian tube.

Signs and Symptoms

Ear ache and deafness are prominent, and it is important to note that there cannot be an abscess in the ear unless there is some deafness. The eardrum becomes red and thickened and bulges outwards as the pus is formed. The pressure of the pus causes pain as it stretches the eardrum. If the eardrum should burst or perforate, due to the excess pressure, pus will come out of the ear, blood stained at first and with relief of the ear ache. It is incorrect to label any ear ache, no matter how severe, as due to infection or abscess in the ear. Infection is the least common cause of ear ache in adults (see Chapter 8).

The ear is for hearing and balance. In disorders of the ear, hearing is often affected resulting in deafness or hardness of hearing. Deafness varies in degree, depending on the nature and severity of the cause, ranging from a slight dullness due to a head cold, to a moderate deafness with catarrhal fluid in the middle ear and to a severe deafness of a person with an acute mastoid infection. Minor degrees of deafness are not usually noticed, but passed off as a dull or blocked feeling in the ear.

It should be noted that there are many other causes for a blocked feeling in the ear than deafness.

There can be no middle ear infection of significance, without some loss of hearing, and this is the most obvious difference from outer ear infection. Only when hearing has returned to normal, has the infection cleared completely.

Treatment

As the person's hearing can be damaged all patients with deafness should start on antibiotics immediately, and this is changed to the appropriate one as necessary, when the ear swabs are reported on, or if there is no clinical response within 36 hours. When there is no ear discharge, warm oily drops may be used to give relief from the pain, but if there is any drainage or discharge, it should be cleaned away at least three times daily, or more often depending on the amount. Cotton buds (Q tips) are too hard, too large, too rough and not absorbent so should never be used in the ear. It is best to use a wisp of cotton wool rolled onto the end of a match stick. For the ear ache, aspirin or paracetamol should be given. Antibiotic drops are useless for acute middle ear infections without drainage or discharge. Other treatment is given to clear any nasal infection or congestion using decongestant nasal sprays or tablets.

DRAINING THE MIDDLE EAR or myringotomy

This operation is to allow drainage of the pus from the middle ear abscess (purulent otitis media) when it fails to respond to other treatment. It is also used to remove the mucous fluid in middle ear catarrh or "glue ear". The operation cuts a 1/4 inch (3-4mm) slit in the eardrum and the fluid is removed with a miniature vacuum cleaner or suction. It is also used in cases of irreversible otitic barotrauma or the "squeeze" to give immediate relief by allowing air in to equalise pressure differences.

Prevention of Middle Ear Abscess

All swimmers should be taught to breathe out through the nose whenever the head enters the water. Water irritates the lining of the nose which then swells up and blocks the nose in part or completely. Thus it is necessary for divers to keep the face mask clear of water. This prevents nasal congestion and infection which could spread up the Eustachian tube. If there is a hole in the eardrum then water should be kept out of the outer ear canal by no swimming or an ear plug can be used temporarily.

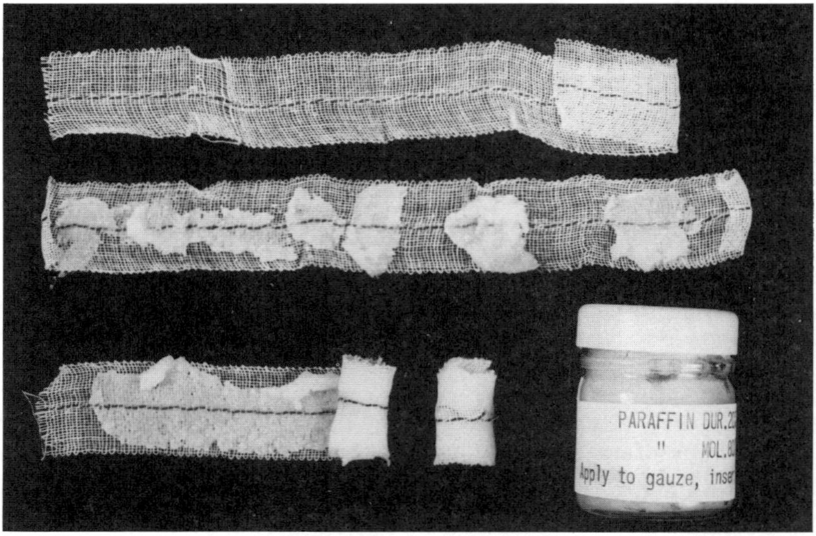

Fig. 6

Paraffin Ear Plug. Six inches of half inch ribbon gauze is coated with a mixture of hard (20%) and soft (80%) paraffin wax and rolled into a cylinder. This can then be worked into the outer ear canal.

For such scuba divers with temporary perforations, if diving below 15 feet (5m) ear plugs cannot be worn as these can be forced into the ear canal. An alternative for depths up to 30 feet (10m), would be a roll of woven 1/2 inch ribbon gauze (Fig. 6) impregnated with hardened paraffin (vaseline) and pushed into the outer opening of the ear canal. Although it could be forced a little into the canal its soft consistency and its loose folding nature allows easy extraction without leaving any behind. If Eustachian tube inflation maintains the air pressure internal to the plug, equal to the external water pressure, the plug would not be pushed inwards. For those divers wishing to go deeper or those who have mastoid cavities, the Queen Hood (Fig.7) should be worn. With this there is a continuous stream of air, coming by way of a tube inserted between the two stages of pressure reduction, and as it is at a pressure greater than the surrounding water pressure, it prevents water from reaching the entire outer ear which fits into the ear cap of the hood. The use of the air for this purpose reduces the time a diver can stay underwater.

MIDDLE EAR CATARRH or secretory otitis media

This is not an infective condition, as there is no congestion nor pus formation. It consists of the collection in the middle ear cavity of a clear, straw-coloured fluid which may be thick, when it is called "glue ear", or it may be thin. It may be the end result of an infection, it may be due to an allergy, but is basically due to Eustachian tube obstruction. It does occur in divers who have had difficulty in equalising air pressure in the middle ear with symptoms of the "squeeze". Any marked negative pressure relative to water pressure results in fluid being sucked from the lining into the middle ear where it can then be seen through the eardrum. This is a form of middle ear barotrauma with deafness.

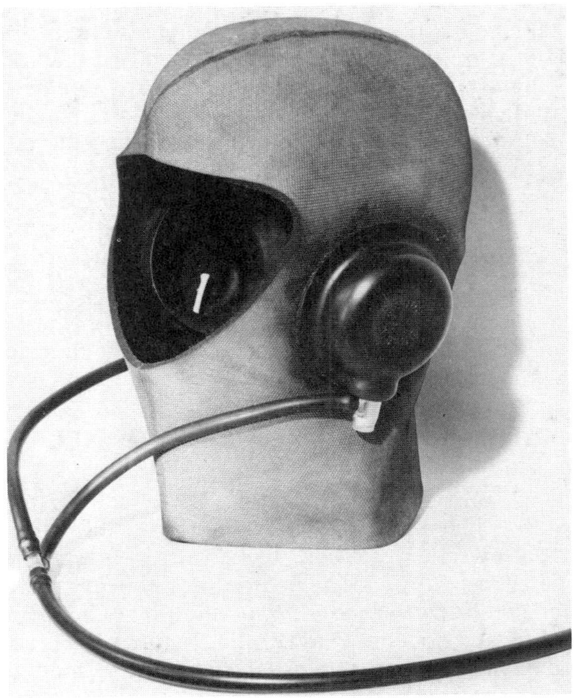

Fig. 7

The Queen Hood. Air enters the ear cup through the fine hole marked with a match stick.

Symptoms

Symptoms are an occasional earache, deafness, sound of bubbles or a feeling of a blocked ear. Signs consist of a waxen appearance of the eardrum through which the fluid may be seen, but often the drum appears normal.

Treatment

Treatment is started by treating any nasal disorder (see Chapter 6). Once the nose is healthy, the middle ear is then inflated to clear the Eustachian tube by Politzerisation (Fig.33p.100) or preferably auto-inflation as for clearing the ears on descent when diving. This should then be done up to 10 times daily until cured.

If this does not relieve the condition, the fluid can be removed by suction, either with a syringe and long needle or after a myringotomy under local anaesthetic. If indicated, a flanged polythene tube or grommet is temporarily left in the incision, as an artificial Eustachian tube which automatically maintains aeration of the middle ear. A tube in the eardrum is not a contraindication to surface swimming. Limited underwater diving can be done provided it is easy to blow air out through the ventilating tube and an effective ear plug is used.

TOYNBEE MANOEUVRE

Pinch the nostrils closed and swallow. The diver should hear a faint click or feel slight pressure in the ears. This pressure is either positive or negative. It was described as a test of Eustachian tube function about 1853 by Joseph Toynbee of London.

VALSALVA MANOEUVRE OR AUTO-INFLATION

For clearing fluid from the middle ear (or equalizing on descent) pinch the nostrils closed and whilst attempting to blow down the nose, carry out a swallow. For practise, the swallow is easier by having a mouthful of water to swallow. The person should hear a hiss, pop or click in the ear or get a feeling of pressure. This should not be carried out unless the nose is healthy and not congested. If the ear does not clear easily leave it for another day and carry on with nasal treatment. A manoeuvre to make it easier to inflate a difficult ear, is to bend the head away from the side that needs clearing. The common factor in divers who develop permanent nerve deafness as a result of diving is the fact that they have had more difficulty in inflating their ears than usual in the water and have descended or ascended rapidly. Persistence in raising the venous pressure in the head by prolonging the "blowing hard" against "pinched" nostrils, say for 10 seconds, may rupture the Round Window membrane in the susceptible person, causing inner ear deafness. Do not hold the nose blocked for more than five seconds at a time. Antonio Valsalva described this manoeuvre in his book published in 1704 for the treatment of draining ears in adults.

Other Methods of "Clearing the Ears"

Of all the procedures learned for underwater swimming "clearing the ears", "equalizing", or inflating the middle ear is the one procedure that is difficult. Valsalva manoeuvre has been described but there are other methods (see Chapter 9). Some divers blow down their noses into the face mask and if this does not have a purge hole, the back pressure is sufficient to force air up the Eustachian tube. Others just give a swallow or wriggle their jaws in a manner they have found will open the Eustachian tube.

Most divers have to hold their nostrils closed or push the rubber of the face mask against their nostrils thus blocking them, and then they blow down their noses. A few have to do this but either blow hard or have to give a swallow as well. If a diver is having difficulty in performing this manoeuvre to clear the ears, he should practise it regularly when he is out of the water. The control of the muscular movements required to open the Eustachian tube can be developed with practise until a few divers can even hold the Eustachian tube open at will.

Case History

Skindiver A.R. a keen member of the Dolphin Club, visited my clinic following a weekend at the Poor Knights Islands, because he had a dull feeling on the left side of his head with his left ear blocked. He had no pain, but the ache he had when he awoke that morning, had passed off after he had been up an hour or so. He had had an uneventful weekend apart from his last dive when he had descended rather rapidly in catching up with his buddy. On this dive he had been able to clear his ears but with some difficulty on the left. He had some left ear pain on descent. His nose had been a bit stuffy from his previous dives and this was aggravated by lying in the sun for the hour previous to the last dive.

His left eardrum was normal looking but through it could be seen a few dark rings of the bubbles in the fluid in the middle ear. His nose was congested, and as there was no sign of any infection, his ear was inflated with the Politzer bag (Fig.33 p.100). He was supplied with an Otrivine nasal spray for the nasal congestion and as he could inflate this ear easily, he was instructed to do it three times daily. When seen a week later, his ear was back to normal and this was confirmed by doing a hearing test with the audiometer. Incidentally he had been out diving again during the weekend between visits.

THE "SQUEEZE" or middle ear barotrauma

As a result of descent, as in an aeroplane or in underwater swimming, there is a change in the external pressure on the ear from low to high. If the Eustachian tube does not allow air into the middle ear, then the outside raised pressure will push the eardrum inwards (Fig. 8) with the dampening down of the movements of the eardrum. This is merely a dulled feeling in the first instance, then pressure, but it is not long before pain is felt due to stretching of the nerve endings in the eardrum. The onset of this pain commonly occurs in the inexperienced diver after a change of pressure of 6 feet (2m) of water. In the more experienced diver the sensations are not so great, partly because of the fact the eardrum has been stretched on many occasions, and partly because of the natural disinclination to take any notice of early warnings. With further stretching of the eardrum pain increases and several effects occur.

The first effect which is commonly seen in divers is that the tissue in the eardrums tears liberating the chemicals that cause inflammation. This causes blood vessels to increase in size so that what are normally too small to see become visible. This effect can last up to one week.

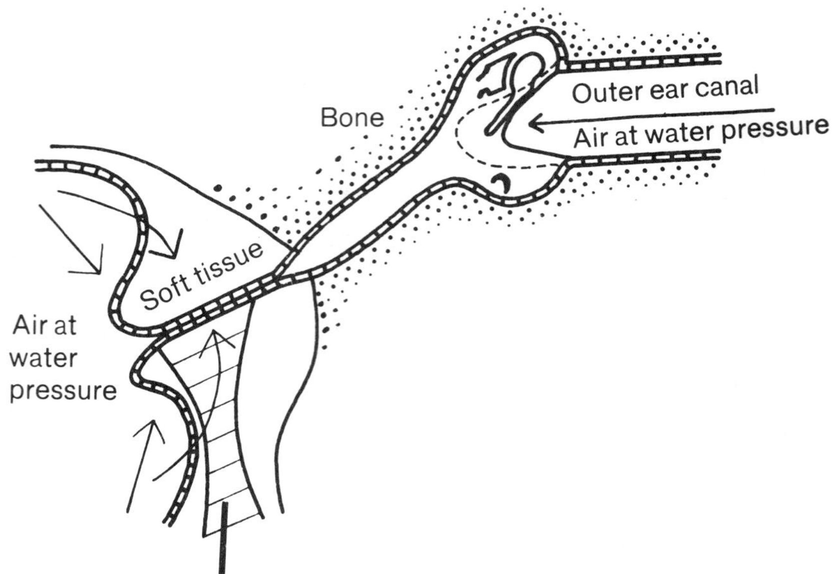

Fig. 8

The Eustachian tube locked closed because the water pressure is greater than the pressure that the breathing-out muscles can produce.

Secondly some of the small blood vessels can be involved in the tissue tearing. These bleed into the substance of the drum causing bruising which can last up to two to three weeks. This bruising is common and indicates poor technique in clearing the ears.

Thirdly, as the eardrum does not stretch far enough to allow the pressure inside the middle ear to equal that on the outside, the developing negative pressure in the middle ear, extracts fluid from the blood vessels into the cavity of the middle ear. Sometimes a small blood vessel may break so that the middle ear fills with blood.

Alternatively, as the outside pressure increases the eardrum may rupture or tear. This occurs sooner if there is any weakness in the drums. It occurs in pressure changes ranging from 6 to 30 feet (2-9m) depending on the strength of the eardrum. If water enters the middle ear through the perforation, dizziness or vertigo occurs. The entry of cold water commonly occurs in depths over 30 feet (9m) and it can be a danger to life.

Repeated episodes of the "squeeze" or barotraumatic incidents will cause repeated collections of fluid in the middle ear and this leads to a weakening of the eardrum. This fluid in the middle ear may also lead to scar tissue and adhesions developing inside the ear resulting in permanent deafness. The repeated small bleeds into the eardrum may result in the thickening of the eardrum seen in some scuba divers, but usually the blood or bruise clears completely leaving no after effects or signs.

Treatment of the "Squeeze"

The diver, if underwater, just needs to come back towards the surface to decrease the pressure differential, and clearing of the ears will be easier. If difficulty in clearing the ear persists then the diver may try the following. With the nose blocked give a quick hard forced expiration. The sudden and explosive increase in pressure may then force air up into the middle ear, equalizing the pressure with the surrounding water pressure. If this or other manoeuvres are not successful it may be necessary for the diver to give up diving for the day. After the dive, if the Eustachian tube remains blocked and the air has gone from the middle ear then a doctor using a syringe and long needle can inject air through the eardrum, or he can cut a hole in the drum as with a myringotomy (page 17). This equalizes the pressure.

With the collection of fluid in the middle ear as described, with Valsalva manoeuvre the diver may be able to clear this fluid away. Inflation can be done up to ten times daily but remember do not hold nose for any longer then five seconds at a time. In those cases where bleeding occurs into the middle ear the diver may suffer continued earache and certainly some deafness. Under those circumstances he should give up diving for that day primarily because of his Eustachian tube obstruction.

Free blood in the middle ear is removed partly down the Eustachian tube and partly by being reabsorbed back through the mucous membrane lining. So treatment to clear the nose and Eustachian tube is given. Permanent changes can occur if the free blood takes more than 5-6 days to disappear and the recovery from such a bleed should be confirmed by a return of the hearing to normal. Only after 5-6 days would a myringotomy and suction removal of the blood be indicated. To prevent any infection which could spread readily in the remaining blood, the myringotomy would be done with strict aseptic precautions.

Prevention

As the diver descends he normally should inflate his ears starting on the surface, by the method which he has found easiest and if any discomfort develops, he should stop descending or ascend a few feet (½-1m) until he can clear his ears. As the pressure differential increases, the soft tissue of the inner end of the Eustachian tube acts as a valve and closes more firmly, causing "locking" (Fig. 8), so that early rather than late inflation as indicated.

Factors preventing the Eustachian tube opening and factors helping it should be examined (see Chapter 2). The technique of inflating or clearing the ears should be practised repeatedly out of the water so that it can be performed easily and frequently when descending, without much concentration or effort.

Case Histories

This is an example of a mistaken idea of a diver who thought he could not inflate his ears. Although he went through the correct motions of clearing his ear, he was often left with a blocked feeling in the right ear. Accordingly he would strain and force more air up the Eustachian tube and it would feel more blocked. So he would relax, ascend until the ear felt right and then descend slowly. His nose, Eustachian tubes and middle ears were the same on each side but he had an impacted right wisdom tooth which caused, even when not swimming, a blocked feeling in the right ear. So I asked, "What depth have you been in?" He was a learner, diving in 6 feet (2m) of water and his instructor had sent him for treatment of a "reluctant Eustachian tube". Excessive clenching of his teeth (see Chapter 8) would bring on the blocked feeling as he descended. As soon as he began to relax and take things more slowly and deliberately, the blocked feeling would pass off. It was due to muscle tension inside the ear. This tension would pass off as he relaxed and took things more slowly and deliberately.

Case H.A., again a beginner was sent by her surgeon father because she said she was having difficulty in clearing her ears. She said that she never heard a "bang" in her ears when she tried to clear them. Her instructor had said for her to listen for the "bang" but all she heard was a gentle click, with relief of the pressure feeling, indicating a better than average Eustachian tube. She was reassured and carried on diving without any trouble.

Case No. 1024 was a 20 year old female who had been diving for about one year. She had had trouble all along with clearing her right ear and she thought she should do something about this problem. She would suffer pain on descent only in the right ear and she would go up and down trying to clear this ear which she would do eventually. However, when she started noting a blocked feeling and deafness on surfacing she was motivated to action. The deafness would pass off after a day or two. She was seen the day after a dive and the Toynbee manoeuvre was only positive on the left. She had minor exostoses (Fig. 9, p. 31) but bubbles could be seen behind her right eardrum. Bubbles of course means a mixture of air and fluid. Treatment was auto-inflation every hour on the hour when awake until the deafness went. She was counselled and coached on this. Once the fluid went she had the right inferior turbinate in the nose, cauterized as a preventive measure. This treatment reduces nasal congestion or "sinus".

Key Points

1. If your hearing is normal then your middle ear has no problems.
2. Abscess in the middle ear is rare in adults.
3. Inflating the middle ear keeps it healthy.
4. Inflating the middle ear prevents the squeeze and middle ear damage.

CHAPTER 4

THE OUTER EAR

This is the part with which everyone is familiar. It consists of the leaflike pinna which juts out from the side of the head and the outer ear canal. The latter is the skin-lined passageway leading into the head, and in which wax may collect. This ear canal is about 1½ inches (4 cms) in length, forms an S-shaped curve and has a blind ending formed by the eardrum (Fig. 4). The thicker skin of the outer third contains hairs to prevent the entry of dust and other particles, and has in it, wax-producing glands. The inner half of the outer ear canal is lined with very thin, easily damaged, hairless skin. This part is in the bone of the skull. At the junction of this bony part with the softer, mobile outer half of the outer ear canal, there is a weak or stress region.

WAX IN THE EAR

This is not a disease, but a normal occurrence. It is produced from modified sebaceous glands in the outer third of the ear canal. Wax is produced continuously in normal ears, the quantity varying from person to person. The normal cleaning mechanism of the outer ear canal is by migration of the top layer of the skin. This layer moves sideways from the eardrum to the outside carrying all surface material to the outside. Wax is thus brought to the outside where it dries, powders and drops off. Or it may be removed with a short fingernail or a face cloth. If too vigorous attempts to remove it are practised, the wax is usually pushed inwards and it may even be pushed down onto the eardrum.

The wax provides waterproofing of the skin to prevent it becoming macerated, like dishwasher's hands. Those mammals which have returned to the sea permanently, have all evolved various mechanisms to keep water out of the ear. The whale has a plug of wax, filling the entire outer ear canal permanently and its consistency prevents it from being washed out, but does not prevent the sound waves in the water from reaching his inner ear. Those swimmers who have it washed out by swimming, are liable to infection of the skin or external otitis. However the S-shape of the outer ear canal does prevent wax washing out in the vast majority of swimmers; let alone the hood of most divers.

It is not always wise to syringe an ear from which there is any discharge, nor an ear that is aching or painful. Those people who prevent the wax from coming out may need it removed by a wax curette or by syringing of the ear. If syringing is done too often, the resulting irritation leads to further poking or rubbing the ear leading to the necessity for further frequent syringing. Collection of wax in the ear is a self-inflicted condition. The diver should be reminded that he should not pick at nor put probes, cotton buds, hairpins or sticks into his ears. Nor should he rub in or around the ears. Poking into the ears is a common habit with those with itchy or blocked ears which is usually an early sign of skin infection or the jaw joint disorder (see Chapter 8).

If there is any trouble with the ears, warm olive oil can be put in and if still troubled, consult your doctor. Eurax (Geigy) anti-itch lotion can be supplied to those with normal looking ears but who poke them because of the itch or irritation. Within five seconds, this lotion will abolish the itch and then there will be no desire to continue poking or rubbing at skin around the ear. To dry the ear, alcohol as in Spirit ear drops (page 103) can be put in. Sometimes after diving or swimming, the water will not drain out of the ear. Spirit eardrops or even two drops of gin or whisky, will lower the surface tension of the water which will then run out and the blocked ear feeling will go.

OUTER EAR INFECTION or external otitis

This is an infection or dermatitis of the skin of the outer ear canal. The two main types of outer ear infection are the localised boil, and the widespread or diffuse external otitis caused either by bacteria, fungi or allergy. Fungal infections are often resistant to treatment and are usually made worse with antibiotic drops. Fungal infections occur more frequently in tropical climates. Hence we hear of Singapore Ear or Caribbean Ear or Coral Ear and other names.

BOIL IN THE EAR

This is like a boil elsewhere, except that as the skin in which it occurs is firmly attached to the underlying cartilage or gristle, there is not the usual great swelling and there is more pain. The inner half of the outer ear canal does not have hairs and so a boil is always situated in the outer third. The bacteria gain entry through the skin, often as a result of some slight skin scratch as a result of sticking objects into the ear or rubbing with the finger in or around the ear.

Treatment

This consists of gently but thoroughly cleaning out the outer ear canal and the insertion of a wick. A wick is 1 inch of ½ inch ribbon gauze or there are more expensive commercial types. Then the black tarry drops of Glycerine and Icthyol are used, two drops every hour while awake until the pain goes. When the pain has gone the wick is removed, the ear cleaned and normal antibiotic eardrops are given as for Diffuse External Otitis.

Further relief of the pain can be given by hot bathing of the ear. Antibiotics by mouth are usually used and on rare occasions the boil may need cutting out.

Hot Bathing

Take a wooden spoon and wrap a face cloth around it. Dip this into a jug of just-boiled water and then hold it up to the ear. When it is cool enough to touch the ear, return it to the water to re-heat it. It is the radiating heat that does the good, so do not touch the ear nor let the water get at the ear. This procedure is soothing, relieves the earache, tends to draw out the infection and is very good for all types of earache.

OUTER EAR INFECTION or diffuse external otitis

This is an infection of the entire skin of the ear canal and may spread outwards onto the pinna. It is a skin disorder. A predisposing cause of outer ear infection is prolonged immersion in the water with the removal of the normal ear wax, then poking the ear with match sticks, hairpins or cotton wool buds (Q tips). The longer the immersion, the greater the removal of the protective wax. It is known to be more common in tropical climates because the warmer water washes out the wax coating quicker, the humidity is greater and there are more airborne infections about. Wax in the ear is a normal protective secretion and its absence increases the chance of developing external otitis. Where present as a thin coating, its removal can only lead to harm, but where present in lumps, it prevents water from draining out freely and can lead to outer ear infection.

Infection from water can be a cause of outer ear infection but it is certainly not so in the case of chlorinated pools. It is highly unlikely in the open sea due to the dilution factor, but it may be a factor in untreated or overloaded pools or where water pollution is high.

Case History

This was about a team of 10 scuba divers employed by the Ministry of Agriculture and Fisheries. To prevent spread of lake weed to the hydroelectric plant downstream they were used to harvest the lake weed by hand. The lake weed was growing in Lake Wanaka in the area fed by the town sewerage outlet. At one time the whole team suffered severe diffuse outer ear infections to the stage where the Public Health Medical Officer was trying to stop the whole programme. He stated that diving in sewerage was a grave risk to health. Lake-bottom sediment samples near the sewerage outflow revealed the same harmful sewerage bacteria as did the ear swabs. The Ministry requested help and the problem was investigated and treated.

Firstly the local enthusiastic doctor was forbidden to poke ear swabs into the ear canal. This can and does damage the very thin delicate skin of the outer ear canal. Secondly the divers were strongly requested to cease rubbing in or around their ears and certainly not to poke corners of face cloths or towels into their ears. Thirdly the divers had Spirit drops (page 30 and 103) put in their ears, once each day after diving. Firstly the left ear, the head, kept over for one minute and then the right ear soaked for one minute. With this routine over 1,000 diver hours were logged without any ear infection. In fact two divers who had outer ear infection before the trip, within two days of the use of Spirit drops, were cured of their ear infections despite diving in the sewerage outlet.

Signs and Symptoms

The main one is the itch or a blocked sensation, but there also may be an ache or discharge from the ear and rarely pain. The discharge may be thin and run out easily, or it may be thick and look like wet blotting paper. Sometimes it dries in flakes or scales which may be difficult to remove from the ear. The skin itself is reddened and sometimes swollen and sometimes very tender to touch.

Treatment

This consists of weekly suction cleaning out the ear and the use of eardrops. These are commonly the antibiotic drops of Neomycin or Soframycin, with the cortico-steroid drugs such as hydrocortisone or dexamethasone. Vosol with 2% glacial acetic acid and 1% hydrocortisone added, or anti-fungal preparations such as Jadit solution or Mycostatin dusting powder are also used. Before putting the drops in the patient should clean out the ear gently. A piece of cotton wool twisted onto a match stick or orange stick can be rotated down into the ear canal. If it comes out moist, do it again with a clean wisp. Cotton buds are too big, too hard, too rough and not absorbent and should never be used in ears. When the condition has subsided it leaves the skin in a weakened condition and the infection can recur. The Vosol is then used weekly for 4 weeks and this tones up the skin. See Chapter 12 and the Drug List for the full prescriptions or ingredients of the various ear drops.

Case History

On examining **V.C. a harbour diver** who spent up to 8 hours daily for most of the year on or under the water, the skin of his outer ear canal had no wax and was macerated. A most likely case to develop skin infection. He said that he had never had any trouble with infection and he put it down to the fact that he never poked anything into his ear. He was quite dogmatic when he said, "I know better than to pick or poke my ears and this is why I never have trouble". He was very right in his knowledge and had no loss of diving time with his contract pipe-laying in the dirty water of Singapore harbour.

Prevention of Outer Ear Infection (External Otitis)

(1) Do not pick at, poke anything into or rub in or around the ear.

(2) Do not use cotton wool buds anywhere near the ear. The ear has its own cleansing mechanism. Cotton wool buds can damage the ear.

(3) Do not pick at the nose from which infection can be carried to the ear.

(4) In susceptible people it may be necessary to continue with treatment with Spirit drops for a longer period. Five percent Glacial Acetic Acid in Surgical Spirit used after diving or weekly can prevent a recurrence.

When there is an established outer ear infection the less water there is in the ear the better, but swimming or diving is allowed provided that the Spirit drops are used after swimming. The normal ear skin is acid and this is maintained by the acetic acid which at a strength of 5% kills bacteria and fungi. The Spirit absorbs water and thus dries out the ear canal.

(5) It is recommended that frequent divers or swimmers should have a yearly medical inspection of ears for the likelihood of exostoses (see later) and for the removal of any excess scales of skin or lumps of wax.

Case Histories

Scuba diver B.R. a keen man, was in despair because of the trouble he was having with his ears. Trouble in clearing them when he dived and all the time developing infections in his ear and having to keep out of the water on the order of his own doctor. He did not keep out of the water because he wanted above all to dive, and so he did. He imported, at some expense, a Queen Hood (Fig. 7) from the States, and this kept his ears away from water and infection free, but utilised too much of his air supply, limiting his time under water. Removal of his adenoids (Fig.20,p.62) and deep X-ray therapy to his Eustachian tubes completely cured his Eustachian tube blockage. His external otitis or outer ear infection, settled within a week with antibiotic-steroid drops (Neomycin-betamethasone). By using the 5% Glacial Acetic Acid in Surgical Spirit as drops before and after diving, he was able to dispense with the Queen Hood. Mind you, he did not give the Queen Hood up, but continued using it without connecting it to his air supply. It looked rather mod! Twenty-four years later he was back to see me as he had developed another external otitis!

There is a famous comedy novel called "Three men in a boat" but this was "Three men and their cotton buds". Within one month these three men had consulted me. They had poked their cotton buds (Q tips) right through their eardrums. They said rather shamefacedly that it was very, very, painful! The perforations or holes in the eardrum all healed without operation.

SWIMMERS NODES or exostoses

Another factor helping to cause external otitis is the presence of exostoses (Fig. 9). An exostosis is a narrowing of the inner half or the bony part of the circular outer ear canal by bony swellings. These localized bony swellings grow slowly over a period of years in response to the irritation of cold water on the skin of the outer ear canal. They are called "swimmers nodes" and are common in the surfing cult.

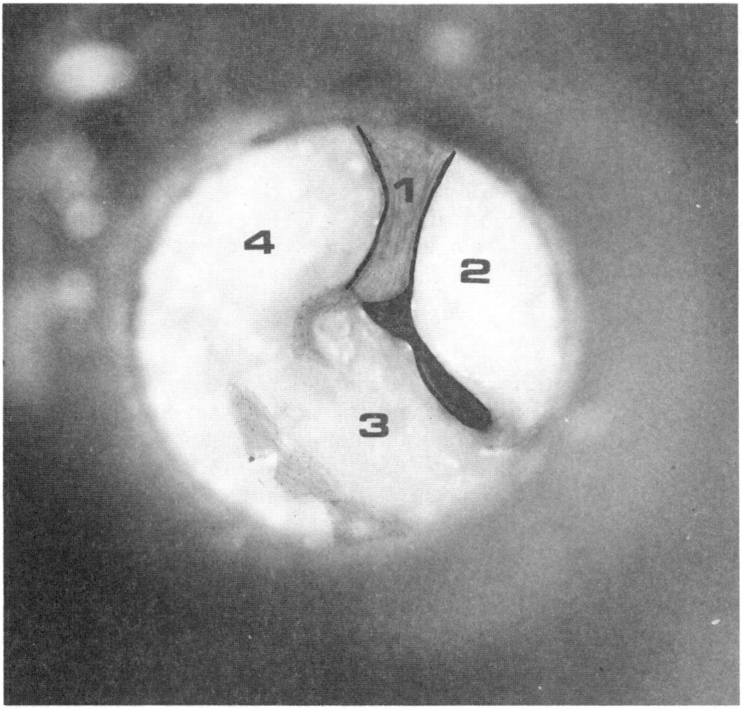

Fig. 9

Large Exostoses. The normally circular outer ear canal is mostly blocked with four rounded bony swellings.

This condition is not related to infection nor is it caused by infection. This blocking of the ear canal, by preventing water from draining out, makes the person susceptible to outer ear infections. The bony swellings continue to grow while there is a continued exposure to cold water, and the temperature that qualifies as cold, is that found in sea water and outdoor unheated swimming pools, in temperate climates.

It is rare in swimmers confined to the Mediterranean and even less so in tropical swimmers. The ear canal may become almost closed with resultant deafness, but at any stage surgery can return the ear canal to normal.

Prevention of Swimmers Nodes (Exostoses)

When the medical examination reveals the early development of, or there is an established exostosis, it is important that measures are taken to prevent the continued growth of the exostosis. If swimming in non-tropical waters a hood should always be worn, or for surface swimmers some form of ear plug should be used (Fig. 6, p. 18). A trial of eight different ear plugs for surface swimmers was carried out. The purpose was to find out which was the most user friendly and most effective at keeping water out of the ear. The variety included expensive custom-made silicone ear plugs, compressible foam, malleable plugs and others. The most effective plug was a piece of cotton wool pushed in a little way with a surface smear of vaseline or soft paraffin. This was also considered to be the most customer friendly and acceptable. However the Paraffin Ear Plug (see Fig 6) is even better and can be re-used many times.

Case History

Case No. 848, a thirty-year old male office worker went diving about Christmas 1983. He said he was a little deaf for several weeks after that dive. He consulted me about six months later because he was wanting to again go diving. He said that his ears were always itchy and scaly and he rubbed them a lot. He had a scaly outer ear infection and some wax. At the time of his last dive the water getting into his ear had wet the scales and wax which had swelled up. This wet wax and scale came against the eardrum and caused some deafness. As it dried and was moved away from the eardrum his hearing came back. His ears were cleaned out, eardrops given and he was asked to stop rubbing his ears. He reported back after an uneventful dive and the outer ear scale and infection had gone.

OUTER EAR SQUEEZE

This is not a serious condition, but can make a person unfit for diving, so it has a considerable nuisance value. It is a similar condition to the "squeeze" of the middle ear but occurring in the outer ear canal. It results in damage to the skin including the eardrum. It occurred in four only out of a series of 650 divers.

Cause

The outer opening of the ear canal can become blocked by the hood especially if it has ear cushions or by the unwise diver wearing ear plugs and diving too deep. On descent, a negative pressure develops in the outer ear canal in relation to the surrounding water and to the middle ear, which has been cleared by auto-inflation. This negative pressure is maintained by the elasticity of the rubber hood blocking water from getting into the outer ear canal. To get sufficient pressure differential to cause trouble, the diver needs to be at a depth of 10-30 feet (3-9m) in comparison to the 6-10 feet (2-3m) for middle ear trouble.

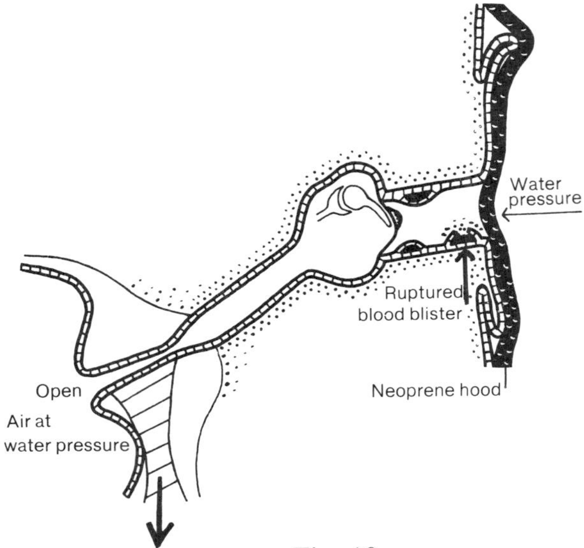

Fig. 10

Outer Ear Squeeze. The Eustachian tube is opening but the outer ear canal is blocked resulting in a low pressure in this canal.

Effect

The eardrum will bulge outwards and cause earache though seldom sufficient to cause a rupture. Because the blood pressure is at normal level plus the pressure of the water, the negative pressure in the ear canal causes fluid and even blood to leak out of the vessels to form blisters (Fig. 10) in the skin lining of the ear canal and in some cases on the eardrum itself. These blisters can break spontaneously giving a tattered look to the eardrum and ear canal with bleeding from the ear. Similar collections of blood can sometimes be seen in the skin of the arms or legs where, as a result of wrinkles when wearing the old dry suit, a similar degree of suction occurs. The wrinkles trap air in which a negative pressure develops.

Treatment

The use of antiseptic ear drops or powder to clear or prevent infection, and as it heals, Spirit drops to dry out the ear. It will quickly heal especially if the ear is kept dry. If it is painful due to the blisters, these may be pricked with a needle to let out the blood. The damage seen should not be confused with a ruptured eardrum, with which it may appear identical when the blood blister on the eardrum bursts to give a tattered look.

Prevention

If it has occurred then it is reasonable to take the precaution of cutting a small hole in the hood over the ear to let in water. This improves hearing out of and under the water and some improvement in clearing the ears is also claimed.

If the ear canal is allowed to fill with water then there is no risk of outer ear squeeze, as pressure equalisation is automatic. Too large a hole will allow free flow of water and at depths the coldness of this water can bring on vertigo or dizziness. The balancing organ is that part of the inner ear nearest to the outer ear canal and cooling of the fluid in it causes a thermal flow. This results in stimulation of the balancing apparatus and a hallucination of movement.

Case History

Scuba diver C.D. was a professional, but had been forbidden to dive by an Ear Nose and Throat specialist who had diagnosed a hole in the eardrum. As he sat in the boat supervising his mate contract pipelaying, he mused on how fate was preventing his wage packet from being as big as it used to be. His mate after a consultation with me, had had his diving headaches relieved, so he recommended C.D. to have a go - "Nothing to lose", and "He might fix you up". So here C.D. was, and he had normal eardrums, no scarring, no thin patches, no thick patches and normal hearing by the audiometer. His story was not typical of a ruptured eardrum nor did he have anything more than a temporary minor hearing loss at the time. His ear did bleed a lot after his last painful dive, during which he said he couldn't clear his ears.

On close questioning, he said he thought he was clearing his ears but the dull pain persisted and he did not usually have trouble equalising pressure. The specialist did remark, "What a mess there is in your ear", so it is most likely that he had an outer ear "squeeze" with broken blood blisters on his eardrum, giving the appearance of a ruptured eardrum. No hearing test was ever performed, then or later, so this unlucky diver had been unnecessarily out of the water for nine months.

Key Points

1. Wax in ears in lumps is abnormal and usually self-inflicted.
2. Cotton wool buds damage the delicate skin of the ear canal.
3. Wax is pushed down into the ear canal by cotton wool buds.
4. Not all swimmers nodes need removing.

CHAPTER 5

THE INNER EAR

The third division of the ear is the inner ear (Fig. 4, p. 14) being the hearing and balance mechanism situated inside the hard bone of the base of the skull. The hearing mechanism, contained in the cochlea, is a device for converting the vibrations of sound into nerve impulses, the actual transducer being the Hair Cells. The Hair Cells do have rows of hairs or fine processes sticking up from the top of the cell. For common purposes they are called Hearing Cells. They are the key for understanding inner ear deafness. The hair-like processes of these cells project into a membrane in the fluid of the cochlea. When these hairs are bent by the vibrations of the membrane, a nerve impulse originates from the body of the cell. The Hearing Cells are very fragile and are easily damaged by blows to the head, excessive noise, oxygen lack, virus infections and loss of or increase of the fluid in which they are bathed. The balancing mechanism with its own type of Hair Cells is contained in a complicated series of circular canals from which the nerve impulses pass to the balancing centre in the base of the brain. The balancing centre is adjacent to the vomiting centre, and one of the further connections of the balancing centre is to the muscles controlling the movements of the eyes. Thus balance problems can cause unwanted, and not realised eye movements.

Deafness due to damage in the inner ear is known as sensorineural or nerve deafness. When the damage affects the balancing apparatus, the symptom is vertigo or dizziness, in which the patient has hallucinations of movement and feels that he or his surrounding are spinning around or are moving back and forwards or up and down. It is important to distinguish between this vertigo and complete loss of orientation with the more non-specific, non-directional giddiness, dizziness or light-headedness associated with mental tension or anxiety states. A decrease in carbon dioxide in the blood caused by over-ventilation or deep breathing can also cause this non-specific dizziness. Mild exposure to carbon monoxide contamination of the compressed air in the tank causes a headache accompanied by nausea and light-headedness or dizziness. In a white-out condition when there is nothing to fix ones' eyes on some divers become dizzy. Some people with this last condition can develop actual spinning dizziness or vertigo.

INNER EAR DEAFNESS

By this is meant lack or loss of hearing, which is rarely total but usually partial and due to damage in the inner ear. It may be so slight that it is not noticed by the patient and is only found by scientific audiometrical investigation. The loss of hearing may be great enough to be obvious to the patient. Commonly, people refuse to acknowledge their slight loss of hearing, blaming it on to other people by accusing them of mumbling. Deafness in scuba divers occurs at one time or another, but permanent deafness is rare and is nerve or sensorineural in type (Fig. 11). Temporary deafness is most often due to a fault in the passage or conduction of the sound waves from the outer ear to the inner ear, known as conductive deafness. Conductive deafness is due to disturbance of the eardrum and middle ear as already described.

Inner ear or sensorineural deafness unless treated immediately and intensively, within a few days, is usually permanent. Sudden sensorineural deafness is recognized as an emergency requiring immediate intensive treatment.

Possible causes of sensorineural deafness in divers are nitrogen bubbles or micro-emboli of the "bends"; or spasm in which the muscle in the walls of the smaller arteries over-contracts to reduce the size of the

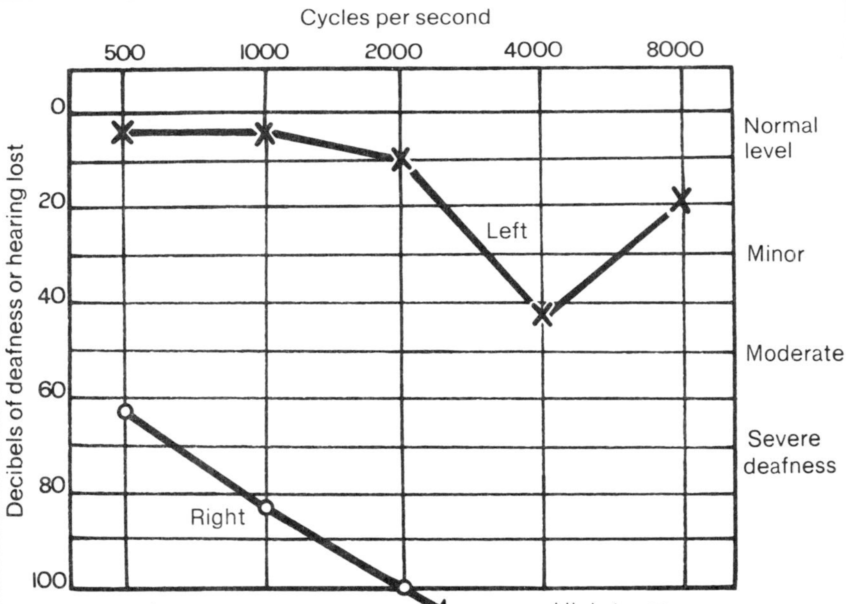

Fig. 11

The Hearing Chart or Audiogram of diver E.T (see page 44). The left hearing (**x - x** in the chart) shows typical noise deafness whilst on the right side **(o-o)** there is a severe deafness with complete loss of the high tones (3, 4, 6 and 8,000 cycles per second). None of the Hearing Cells for these frequencies have survived. He had no treatment for his ruptured Round Window membrane.

vessel and thus reduces the blood flow to a level which deprives the Hearing Cells of their oxygen. This spasm or shrinking of the blood vessels is caused by stress, emotional or physical, and by states of fear or fright. Sudden shock waves as occur in underwater explosions can cause damage in the inner ear just as exposure to loud noise does.

Prolonged and forceful performance of Valsalva Manoeuvre as in clearing the ears causes a transient rise in blood pressure but a more persistent rise in the pressure of the cerebro-spinal fluid surrounding the brain. There is a canal connecting this cerebro-spinal fluid to the inner ear and in some people this canal is large and allows a free flow of cerebro-spinal fluid into the inner ear. With a forceful auto-inflation or Valsalva Manoeuvre, the transmitted pressure increase is sufficient to rupture the thin Round Window membrane (Fig. 12) separating the inner and middle ears. This results in sudden deafness variable in extent with or without some loss of balance or vertigo (see Case History Chapter 10).

The Causes of Inner Ear Deafness

Micro-emboli occur in those cases of "Bends" or Decompression Illness in which the nitrogen gas bubble or other mechanism has blocked one or more of the inner ear blood vessels. Variations in the size and number of blood vessels, between individuals and from one side to the other in the same person, is an important factor in determining which side is affected and by how much. Often an abnormality will only develop as a result of a fine interplay of several variables which singly have no effect at all. The cause of "bends" is well known, but it may occur in instances where the diver has followed the usual rules for decompression, either the tables used by the Royal Navy or the U.S. Navy, or by the diver's decompression meter (DCP meter). By using the Doppler effect, bubbles have been detected in the venous blood stream of divers who should have been quite safe according to Naval tables, so micro-emboli should be considered in all cases. DCP meters are not always reliable especially if used in excess of 100 feet, or if multiple dives on the same day are carried out.

Rupture of the ⅛" (3 mm) diameter Round Window membrane or of the Oval Window ligament allows the inner ear fluid to escape and this can result in reversible damage to the delicate Hearing Cells. Deafness is present with a ruptured Round Window membrane, but rarely there is no hearing loss.

Of 59 cases of sensorineural deafness, Round and Oval Window ruptures only accounted for 9 cases. The other 50 cases or 85% were due to rupture of a membrane inside the inner ear. These cases are similar to Round Window rupture but have minimal vertigo or balance upset. With early and correct treatment these other cases mostly make full recovery. For those who wish to know, it is Reissner's membrane or the Vestibular membrane that does break allowing two different fluids to intermingle causing poisoning of the Hearing Cells.

Symptoms

The onset of a loud ringing or hissing noise in the ear with a blocked feeling or deafness. The deafness is not usually noted immediately, because underwater hearing is negligible anyway. It may be noted early or not for several days, and is usually one sided. Vertigo, with its loss of balance factor, described as the "feeling of being drunk", usually occurs showing involvement of the balancing apparatus in addition to the hearing mechanism.

If questioned, most cases will admit to having had more difficulty than usual in clearing their ears and they may or may not have been suffering from a minor nasal infection or cold. It is this difficulty in clearing the ears, which by necessitating prolonged and forceful expiration against the blocked nose, causes the rupture of the thin Round Window membrane. Other factors that have been recorded in these divers are fast descent or fast ascent. The ear itself has a blocked or dead feeling, giving a sensation of fullness in the ear or that there is a pressure building up in the ear. There may be discomfort with loud sounds which may appear distorted, and with some people the same sound is heard differently in each ear. Although the symptoms of Round Window rupture are usually more severe than Riessner's Membrane rupture, this is not always true.

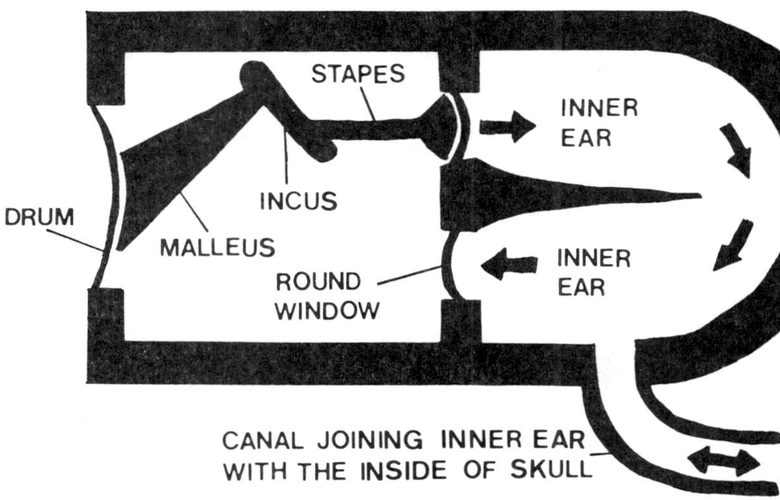

Fig. 12

A diagram of the middle and inner ear showing the connection across the middle ear of the 3 little bones. Note the canal joining the fluid (c.s.f) circulating around the brain to the inner ear fluid (perilymph).

Signs

A loss of hearing, especially in the high tones, but sometimes the whole (Fig. 11) of the frequency range of hearing. The diver at first, if the balancing organ is affected, will have uncontrollable eye movements affecting focusing (spontaneous nystagmus). He may have a staggering gait with a tendency to move or fall to the side of the deranged ear. In severe cases, vomiting may occur and it is needless to say how this may threaten survival underwater. These balancing apparatus signs and symptoms may be present for 1 to 20 days but usually cease. The eardrum usually appears the same as the unaffected ear, and any signs would be coincidental and due to the actual dive or to possible barotrauma or the "squeeze". The absolute sign for a Ruptured Round or Oval Window membrane is seen at time of operation. Fluid can be seen coming through the membrane.

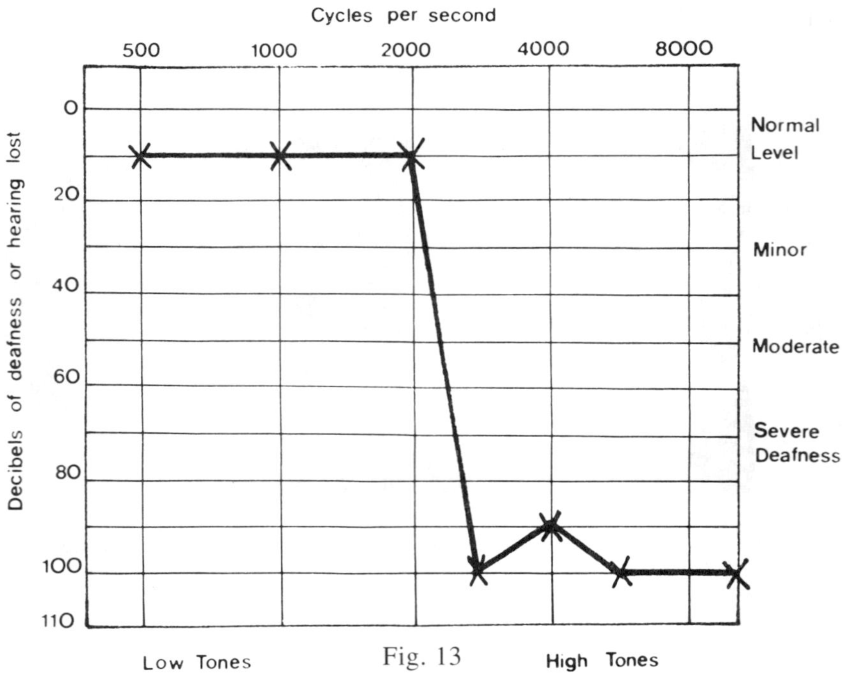

Fig. 13

The Hearing Chart or Audiogram of Case 34 with a ruptured membrane inside the cochlea. Normal low tones (500, 1000, 2000 c.p.s.) and a flat loss of the high tones (3000-8000 c.p.s.).

Other Similar Conditions

Some doctors diagnose this condition as a virus infection of the inner ear by saying that the trouble is coincidental with the swimming. Others have labelled it as Meniere's disease which rarely affects a person with the mental stability required of an underwater devotee, nor does it come on with underwater diving. Some still blame it on a blocked Eustachian tube and give the diver a nasal spray and Sudafed tablets.

Treatment

As the region damaged is in the vicinity of the Hearing Cells, and as the prime need is oxygen, then the treatment of choice is to put the diver in a Recompression chamber and to give him as near as possible 100% oxygen at 1.5 atmospheres. Recompression will remove the micro-emboli of any possible "bends" or Decompression Illness.The extra oxygen dissolved in the blood and eventually in the fluids between the cells, will revive any Hearing Cells that it reaches. Such facilities are available in major centres or wherever there are Navy Diving Schools or Dive ships. With the "Bends" recovery is quick, and this is then followed by slow decompression based on symptomatic recovery; that is, if any symptoms re-appear the pressure is raised again until they disappear and then depressurisation continues. As with other causes of sudden deafness more treatment is needed.

Case History

Case No. 357 was a professional diver who had been doing a helmet dive at 125 feet (40m) for 86 minutes. About 90 minutes after the dive, and this is a delay typical of the bends, he suffered vertigo with nausea and vomiting. He was recompressed and rapidly recovered. His hearing was not affected. This was a case of a the "Bends" of the Balancing Apparatus of the inner ear.

As a second choice for Decompression Illness and first choice for other causes of sudden deafness straight oxygen can be given but the ordinary methods are inadequate. As First Aid, using the diver's own regulator attached to a cylinder of oxygen, although wasteful, does give a high concentration. The ordinary plastic face mask can deliver up to 40% oxygen, but if the head and shoulders of the patient (Fig. 14) are then enclosed in a polythene bag, with a flow of 8 litres per minute, up to 90% oxygen can be obtained. Carbon dioxide collects inside the polythene bag and this further increases the rate of breathing and the oxygen in the blood.

This method of oxygen administration can also be used inside the Recompression chamber so that the diver alone is on oxygen with a consequent saving in cost. The oxygen treatment should be repeated over a period of several days based on symptomatic recovery and a minimum one hour twice daily is recommended. When available Carbogen can be used outside of a chamber instead of pure oxygen. This is oxygen with 5% carbon dioxide and is given for 20 minutes every two hours.

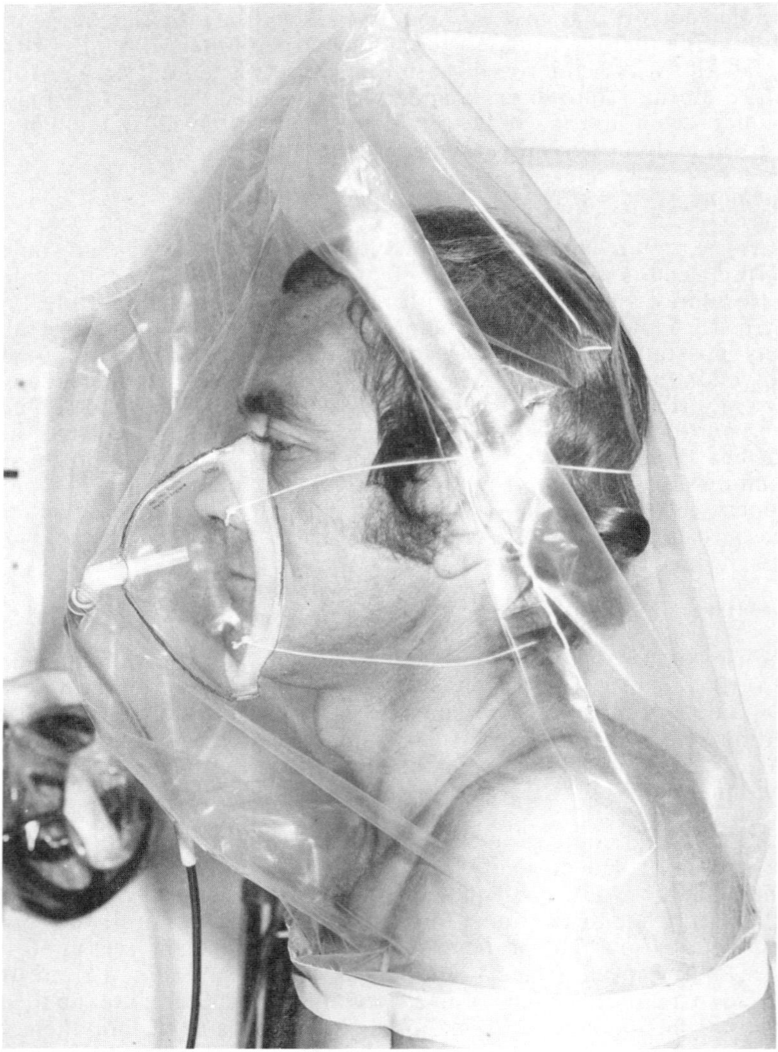

Fig. 14

Patient having Oxygen Therapy. A large polythene bag, placed over the head, is fastened around the shoulders and upper chest with sticky tape. Oxygen flow through a plastic oxygen face mask is at 8 litres per minute.

At the same time there are drugs routinely used for any case of sudden sensorineural deafness. These are Prednisone, Oxypentifylline (Trental) and Betahistine Hydrochloride (Serc). This treatment is usually given in hospital with complete bed rest to reduce intracranial pressures. For divers it is usual to give this treatment for at least three days as a means of separating Round or Oval Window ruptures from Reissner's Membrane rupture inside the cochlea. If the hearing becomes worse or the dizziness persists operation is indicated to inspect for and to repair the Round and/ or Oval Window ruptures.

In addition to treating the deafness, anti-emetic or seasickness drugs may be necessary for the vertigo. Trilafon 2 mg by mouth, by injection or by suppository, Avomine, Dramamine or Marzine, may be given as often as necessary to control the vertigo and vomiting. Other drugs are Buccastem and Scopaderm patches. Because of the disability that deafness causes it is obvious that specialist treatment is needed and it is required within the first week if there is to be any chance of complete recovery.

Case History

After a Sunday dive which was apparently uneventful **Diver J.G.** noticed a blocked sensation in his right ear, with some ringing sound. He visited his family doctor on Monday, who with minimal examination wrongly diagnosed a blocked Eustachian tube and gave him a nasal spray and Sudomyl (Sudafed) tablets. When seen at my office after a week with no improvement, he was put onto treatment for sudden inner ear deafness because there was inner ear deafness present and the Eustachian tubes and middle ear were normal. He was lucky to make a full recovery. The important thing to note is that there are many different causes for a blocked feeling in the ear and a blocked Eustachian tube is not one of them. The tube is normally blocked and only opens with a swallow or by other manoeuvres. This diver had a ruptured cochlear (Reissner's) membrane which can spontaneously heal with no residual deafness nor dizziness.

The Round Window and the Oval Window membranes can be repaired by a relatively simple operation with recovery of the hearing loss, provided that it is carried out within two weeks. At a later date recovery of balance is usually assured but the deafness may persist.

Prevention

The diver should have a wide margin of safety in his decompression staging to prevent the "Bends" from developing. The occurrence of sudden nerve deafness is another reason for not diving when there is any cause of suspect that the Eustachian tube will require prolonged and forceful Valsalva Manoeuvre. When blowing down a blocked nose never hold the nose blocked for more than 5 seconds. Stop, have a breath and try again. The diver should ensure that both descent and ascent are unhurried. If he does develop the symptoms urgent treatment is required. Do not delay or be put off.

Case History

Scuba diver E.T. was diving at about 30 feet (10 metres) off Kapiti Island. He developed a rather loud ringing noise in his ears accompanied by a slight dizziness. He surfaced slowly and found a rock to climb on to. He tried to stand up but kept falling over. The ringing noise in the head persisted and he found it difficult to focus his eyes, whilst the rocks he was sitting on felt as though they were rocking up and down. His right ear felt as though it had a pressure in it, as if at any moment it would go "pop" and everything would come right. After about two hours sitting, he began to feel worse and started vomiting every fifteen minutes. Everything around him appeared to be spinning, and the ringing changed to a hissing. With his unsteadiness and his vomiting and retching he felt awful. His friends helped him back to the mainland where he was seen by the emergency doctor. The diagnosis of Meniere's Disease was made because the doctor discovered some deafness. The emergency doctor gave him some pills to stop vomiting, but they had no effect and the vomiting went on for a further three days. It was ten days before he could stand without holding on to something and it was then he realised that he could not hear with his right ear. Right ears are affected twice as often as left ears for ruptured Round Window membranes.

The diver, a panel-beater by trade called at my office on his way to a weekend at the Poor Knights Islands, ten months later, when I asked him some questions and examined him. He admitted to having had a slight cold on the day of his Kapiti Island dive and had used a nasal spray. Despite this, he had great difficulty in clearing his ears whereas this was normally performed easily by blowing into his face mask. His hearing test or audiogram (Fig. 11) showed some left deafness at the 4,000 cycle per second note, typical of noise deafness, but he had no useful hearing remaining in his right ear. His noise deafness had occurred, unrealized by him, during his time as a panel beater and was, before the fateful dive, present in both ears. There was no doubt that his right deafness was due to rupture of the Round Window membrane. It is likely that surgical treatment would have restored his hearing loss, if the right diagnosis had been made.

VERTIGO

There are various forms of dizziness which occur underwater. Overbreathing which lowers blood carbon dioxide can cause dizziness. The balancing centre is affected and dizziness occurs. Any anxiety state or mental tension can develop into a feeling of insecurity in the hostile underwater environment, giving a vague sense of imbalance or light-headedness. True vertigo and loss of orientation can occur with some divers when they cannot see the bottom or when they are in the "whiteout" state of white water. This is a visual disorder, due to there being nothing to fix the eyes on. Under such circumstances, the diver should look for his bubbles and follow them up or close his eyes, as in searching for non-existent visual information, he will tend to disregard other postural cues. Vertigo, mild to severe, can occur along with any of the disorders of the hearing mechanism leading to deafness, already described. See Chapter 10 for a more detailed account of vertigo.

ALTERNOBARIC VERTIGO

This type of vertigo due to alternating (alterno-) pressure (-baric) changes, occurs on its own without any effects on the cochlea or hearing apparatus. It is quite a common occurrence in scuba divers and has also been described in Air Force pilots.

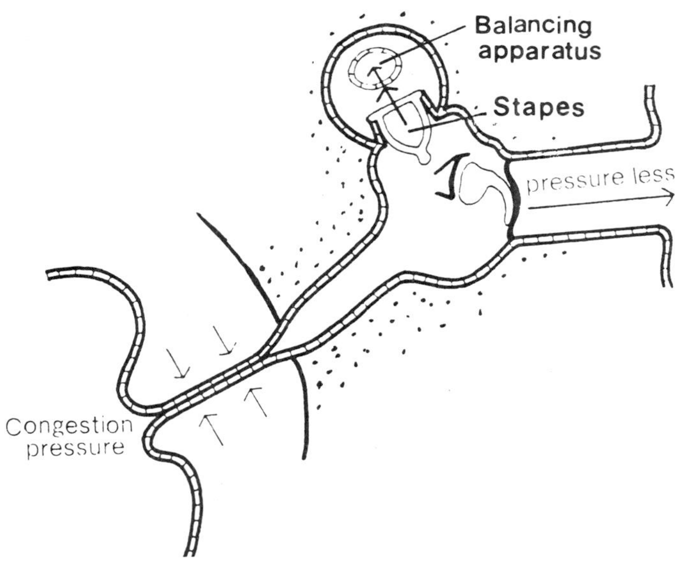

Fig. 15

The "Reverse Squeeze". In ascent, with the Eustachian tube blocked, the expanding air pushes out the eardrum. The pressure also pushes the stirrup bone (stapes) in the oval window, into the inner ear where it irritates the nearest part of the balancing apparatus causing vertigo.

Typically, it occurs during a diving ascent and the diver will, because of the loss of orientation, lose all sense of direction and may move off the vertical ascent into a horizontal position. Eustachian tube blockage causes the trouble with a build-up of pressure in the middle ear ("reverse squeeze") on ascent. It can be due to a too rapid ascent with a blocked Eustachian tube, or a forceful Valsalva manoeuvre in an attempt to equalise pressure in the middle ear but causing overinflation. As the pressure is building up in the middle ear the diver is sometimes aware of it and can slow down or stop his ascent until the air escapes out of the middle ear through the Eustachian tube, often with a hiss. The vertigo is caused by the stapes bone being pushed into the inner ear where it pushes against and irritates one part of the balancing apparatus (Fig. 15). The unexpected arrival of nerve impulses into the brain from the irritated balancing apparatus gives rise to the hallucinatory effect of rotation (see Chapter 10).

Symptoms and Signs

There is severe loss of orientation with apparent rotary movement of the surroundings or vertigo, often described as dizziness. A sudden unwanted change in direction or posture especially if ascending, is noticed by other divers. The feeling of disorientation produces feelings of panic in the diver but usually the surface is reached. In minor cases the diver, if he can see his buddy, will notice a shimmer effect with his buddy seeming to be moving up and down or side to side.

Treatment

As the vertigo is usually overcome, the diver survives, and in future ensures that his ascents are not rapid and that he takes care not to dive when his Eustachian tubes are difficult to clear. If there is any warning that the dizziness is coming on, to go back down prevents the onset. However, most sufferers are keen to get out of the water so continue upwards. Treatment can be given to increase the patency of the Eustachian tubes to allow air to escape easily from the middle ear on ascent, thus preventing the "reverse squeeze" (Fig. 15). If one ear is blocking then the surfacing diver can bend his head away from the affected ear and the pull of the tissues on that side of the neck allows the Eustachian tube to open more readily. The diver can also swallow or swallow while blocking the nose (Toynbee Manoeuvre) or even try to open the Eustachian tube by trying to clear the ears, but gently.

Case Histories

Scuba diver D.S. often accompanied her boyfriend when diving off Tutukaka. She had never had any trouble with clearing her ears and quite enjoyed the diving. After stooging around in depths ranging from 65 to 75 feet (20-24m), time came to return to the surface and when half-way up she suddenly began to lose all sense of direction,and thought she was going to pass out. She doesn't really remember much more until she recovered on the deck of the launch. Her boyfriend "buddy-diver" was trailing her up when he noticed that she had veered off the vertical and started moving horizontally. Her eyes seemed to be flickering in all directions. He seized her and swam vigorously upwards to the surface in quick time. The other divers realized something was wrong and all helped to lift her out of the water into the launch where her dizziness or vertigo slowly subsided. All's well that ends well, but she received quite a shock. However, she is back again diving and makes sure that her ascents are slow and deliberate.

A **35 year old** salt-water marine-land owner from Townsville, Australia, had noticed since 1960 that on ascending from most dives, he developed a pressure feeling in his left ear and then suffered from dizziness. This did not have any rotational or spinning factors. He described it as perhaps a loss of balance, a vague loss of direction with a feeling as though he were falling forwards. One feature which seemed rather constant was his statement that when he could fix his eyes on some object, such as another diver, the object shimmered or appeared to be going up and down. The dizziness was associated with a pressure feeling in his ear and was made worse when he inflated his ears by Valsalva manoeuvre. It occurred more constantly when his nose was stuffy of blocked. He had a crooked nasal septum and small adenoids.

The 29 year old harbour-board **diver E.U.** had been diving constantly for three years and he said "seven days a week". For the past two years he increasingly suffered dizziness on ascending, usually from 30 feet (9m) but even from ten feet (3m). The severity of dizziness being directly related to the depth at which he would be working. In coming up from 30 feet (9m)he would feel the pressure building up in his left ear after five feet (1½m) of ascent, but by ten feet (3m) severe dizziness would begin. He had no rotational factor, subjective or objective, but would feel helpless, and "really seasick" with nausea but had never vomited.

Sometimes he would swim straight to the top as fast as he could and get out of the water "because I usually have 30 seconds before it really hits me. There is a terrific build-up in my ears by this time. My ears will clear with a hiss while sitting on the deck and there are no after effects". He found it hard to focus, and if he tried to walk he would tend to fall forward or stagger from side to side. Usually the dizziness occurred before surfacing but occasionally it would come on up to five minutes after getting out of the water. He had pus in his left sinuses which required a radical operation before it was cured. Then he was no longer having dizziness on ascending. He reported back ten years later with a recurrence of his dizziness but this time he had the spinning sensation of vertigo.

MASTOID CAVITIES

When a person has had a mastoid operation his outer ear canal and the mastoid region becomes one large outer ear cavity, which ends internally at that part of the inner ear containing the balancing apparatus. If cold water gets into his enlarged outer ear canal, because it cools the balancing apparatus, vertigo and loss of balance occurs. People with mastoid cavities should be advised against underwater swimming but if they cannot be dissuaded they should wear a Queen's Hood and confine their diving to depths not greater than 30 feet (9m). There is an operation to fill in the mastoid cavity and return the ear to a near normal state which could help such divers.

Key Points

1. Do not hold your nose blocked for more than 5 seconds.
2. Sudden deafness needs immediate treatment.
3. Vertigo is dizziness plus spinning or rotation.
4. Many inner ear problems are caused by fast descent or fast ascent.

CHAPTER 6

THE NOSE AND SINUSES

As far as safe and pleasant diving is concerned, the state of the nose plays an important part. Any defect of the nose is liable to affect the Eustachian tube, and makes clearing the ears or inflating the middle ear, less easy or even impossible. Accordingly, the commonest medical aid used by divers is some form of nasal decongestant and the commonest cause of trouble in diving is the Common Cold. The complaints to deal with then come under three main headings: conditions causing blockage of the nose, sinusitis and nose bleeds, noting that "sinus" trouble is a modern name for nasal catarrh. All the symptoms of "sinus" trouble, a name invented by folk medicine, are related to the nose and the jaw joint (see Chapter 11). This problem is often a stress related disorder.

DEVIATED NASAL SEPTUM

Instead of being straight and lying as a vertical partition in the midline separating the left side of the nose from the right, the septum often becomes bent and buckled to one or both sides. The resultant defect is known as a deviated nasal septum (Fig. 16). If the defect is severe enough, it prevents the person from breathing freely through the nose.

Signs and Symptoms

The main fault is a blockage of the nose, usually worse on one side, often accompanied by one or more of the following: forehead headaches, frequent colds, attacks of facial pain or sinusitis. The septum can be seen to be mis-shapen and displaced to one side.

Complications

The presence of a deviated septum can prevent the cure of sinusitis, frontal or maxillary. Many other disabilities can occur as a result of the mouth breathing caused, such as chronic sore or dry throats, hoarseness, Eustachian tube obstruction and a chronic ear infection. It is often the cause for prolonged colds.

Treatment

The blockage can be relieved temporarily by the use of a decongestant nasal spray when the blockage causes trouble. Nasal sprays are only of use for minor displacements of the septum and only for short periods. More permanent surgical treatment is an operation known as a Submucous Resection of the Nasal Septum (SMR) or Septoplasty.

THE COMMON COLD

This is a definite, separate, infectious disease which affects the nose and throat, begins in a set way, and runs a set course of at least 3-4 days and either clears up completely or goes on to a catarrhal infected state and on to other complications. It is caused by its own virus and is often called "the flu" by people who would like to think they are more ill than they are.

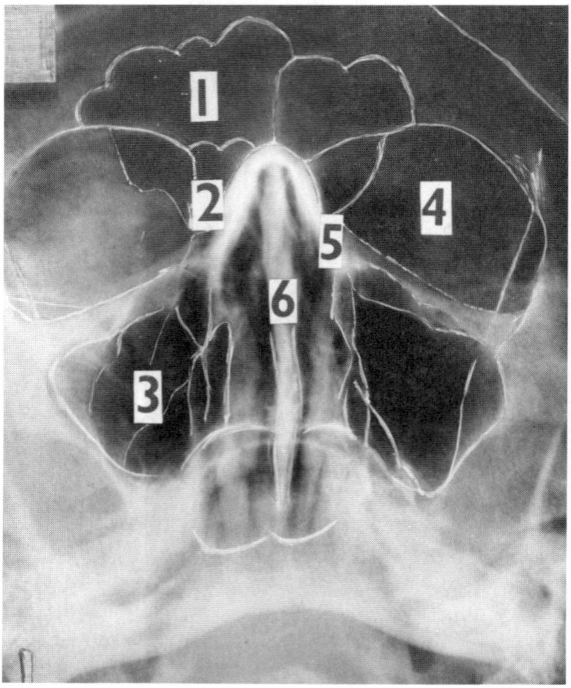

Fig. 16

X-ray of normal sinuses. Frontal (1), Ethmoidal (2), Maxillary sinuses (3). The orbit is (4), the right nasal bone (5) and the septum (6) is deviated to the right of the midline. Normal sinus X-ray is the usual state in sufferers from so-called "sinus" trouble.

Signs and Symptoms

There is a burning at first at the back of the nose and sometimes a sore throat. The patient may not feel too well, be unduly sensitive to cold and may have a small rise in temperature. Next comes a thin watery discharge from the nose, which itself may or may not be blocked. The discharge may clear or become infected, in which case it becomes thicker and coloured. Sneezing which spreads infection, is present in the early stage.

Treatment

There is no useful remedy which can cure a cold, or directly kill the virus responsible for the infection, but treatment is aimed at relieving the symptoms and preventing complications. On the whole, most cases are not treated, but the sufferer puts up with the inconveniences and, in due course, the vast majority of colds clear spontaneously. Simple measures can make a cold less troublesome. Aspirin is taken for any fever, sore throat or face ache. Antihistamines and "cold" remedies from the chemist are helpful in some people. Extra rest in bed is obtained by early retiring. Simple steam inhalations to relieve any nasal burning or pain are very useful, while above all, the patient should be kept warm and well wrapped when he ventures outside. That is, there should be no swimming nor diving.

Nose Blowing

After the first two or three days, in order to clear the nose and dry up the nasal discharge, tablets or capsules to clear the nose or nasal sprays are helpful. In all cases the nose should be kept as clean as possible by adequate and careful blowing. For this, the fully opened handkerchief or disposable tissue is held loosely over the nose around the nostrils, that is, against the cheeks on either side, and the upper lip below. On no account is the tip of the nose to be gripped nor are the nostrils to be pinched. If this is done, the free flow of air is impeded, and the throat and nasal air pressure is raised. The raised pressure can force the infected mucus up the Eustachian tube to cause an abscess in the ear or otitis media, or in through the opening of the sinuses to cause sinusitis. With the handkerchief loosely held in position, the patient blows sharply down his nose, the air coming through both nostrils. If any discharge comes away, the procedure is repeated for as many times as it is necessary to completely clear the nose. The efficiency of the blow depends on the rush of air through the nose, its speed rather than its pressure, and the nose blowing should be continued until no more mucus is produced and there are no longer any moist sounds in the nose. Adults may block one nostril to increase the speed of the air through the other.

NASAL ALLERGY or Allergic Rhinitis

When an attack of sneezing, nasal blockage and watery nasal discharge occurs as result of irritation of the nasal mucosa by a specific irritant, such as a pollen or house dust, the person is said to be suffering fromallergic rhinitis. If the irritant is a pollen, which occurs in spring, then the attacks are known as Hay Fever. If the attacks occur the whole year round due to house dust, animal drandraff or kapok etc, then it is known as a Perennial Rhinitis. Allergy to food, such as milk and those items made from or including milk, is also a common cause of nasal allergy.

Symptoms

Sneezing in attacks when exposed to the irritant is the most obvious symptom, but varying nasal blockage and runny nose also occur. There may be headaches, face-aches or a dry cough.

Complications of Nasal Allergy

The nasal congestion blocks the Eustachian tube making clearing of the ears more difficult. In the midst of an attack, due to swelling of the mucous linings of the nose, the drainage from the sinuses is interfered with, and an infective sinusitis may develop. As the mucosa swells,

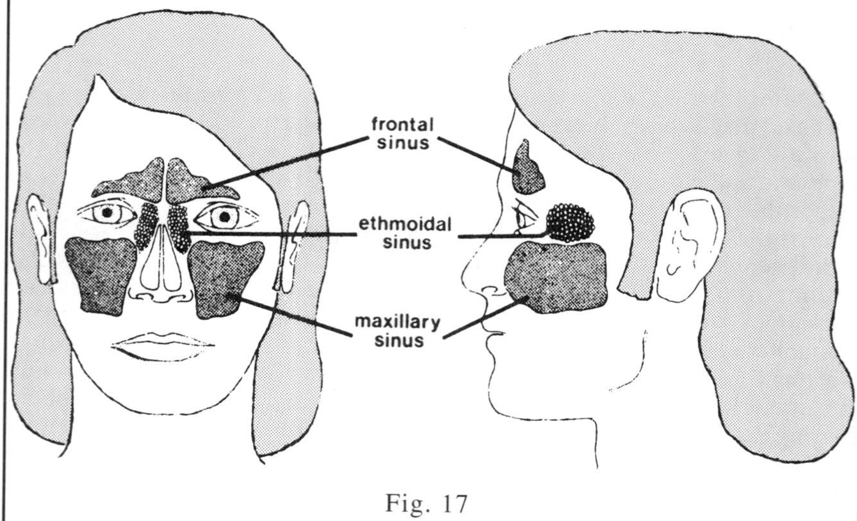

Fig. 17

The situation of the various nasal sinuses superimposed on the surface.

certain areas do so more quickly and they hang downwards, creating small localized rounded lumps or nasal polyps. When, due to reaching a certain size, the polyp becomes permanent, it blocks the nose. An operation to remove the polyps is usually necessary.

"SINUS" or nasal catarrh

The majority of people who suffer attacks of nasal blockage and post-nasal drip, and who sneeze a little more than average, perhaps suffer from aches in the forehead or under the eyes have "sinus" trouble. The cause of their trouble is instability of the control of the amount of blood flowing through the large arteries and veins in the nasal mucous membrane (see nasal congestion, page 10). In those with this tendency to "sinus", the precipitating factor is usually some form of stress such as an emotional crisis, overwork, tiredness or one of the many other varied stress factors, causing nasal congestion and increased mucous production. This used to be known as nasal catarrh and if associated with headaches or face-ache it is now commonly known as "sinus" trouble. Note that the sinuses (Fig. 16) are not infected and "sinus" trouble is not the same as sinusitis, but a misnomer for nasal catarrh. "Sinus" trouble has become well-entrenched in folk medicine since the 1920's, and as such is not used in medical textbooks.

Treatment of Rhinitis and "Sinus"

Sudafed, antihistamines or other nasal decongestants from the chemist in the appropriate doses, may be administered during attacks. Continued use of nasal drops can only cause harm, because most of them are nasal irritants, certain drops being more harmful than others. For short-term treatments such as of a cold or an attack of otitis media, the nasal drops or sprays of "Otrivine", "Drixine" or "Atamol" are worthwhile, but should not be used for longer than two weeks at a time. There are simple operative procedures on the nose (not the sinuses) that give more permanent relief for "sinus" trouble and its effect of causing difficulty in clearing the ears. One of these is cautery or heat treatment. Face-ache and headaches are treated, believe it or not, by jaw muscle stretching exercises as described in Chapter 11.

NOSE BLEEDS

Causes of Nose Bleeds

Often there is a tendency for the nose to bleed due to a large blood vessel being close to the surface at the front end of the nasal septum. Such vessels are abnormal and fragile. In such a condition, only mild injury, raised local blood pressure or local nasal congestion is necessary to rupture the vessel and start it bleeding through the surface. In the vast majority of cases the problem is simple and easy to treat. However lay people tend to get over-excited and worry needlessly. It is not uncommon in divers due to pressure changes.

Treatment

The various methods will be described in order of simplicity. Any one of the procedures may be performed, depending on the cause of the bleed and the state of the patient. Permanent treatment is simple and effective.

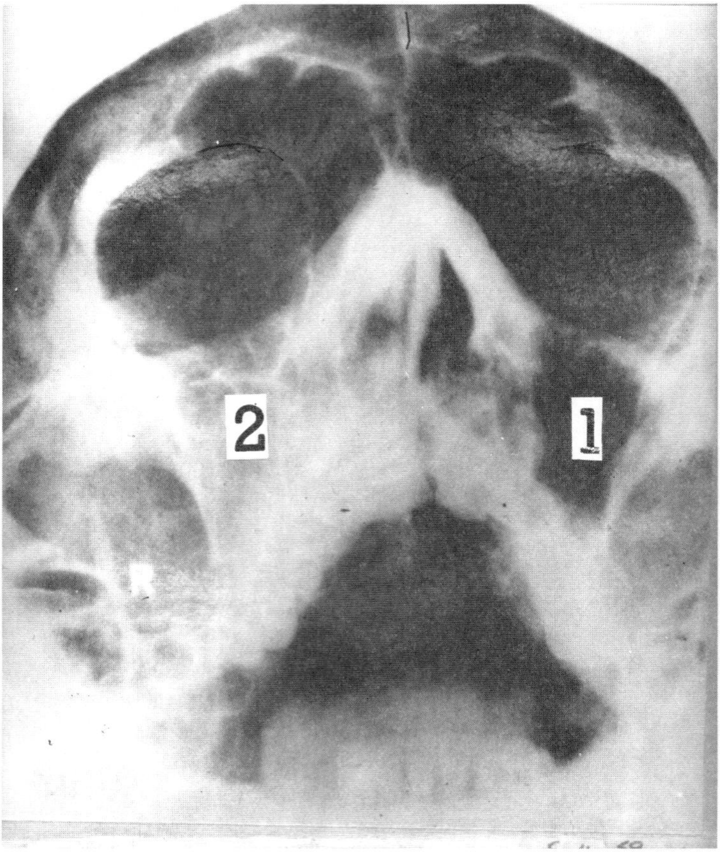

Fig. 18

X-ray of sinusitis. Note the black nature of the air-filled left maxillary sinus (1) when compared with the pus-filled right sinus (2).

Spontaneous Stoppage

In this case, after the loss of a small amount of blood, the bleeding ceases. The cessation is due to nature's methods of stopping bleeding, that is normal blood clotting.

Factors to Encourage Spontaneous Stoppage

Cease nose blowing and wiping as this removes the clot as it is forming over the open blood vessel. Pinch, with thumb and index finger, the whole of the fleshy part of the tip of the nose including, if possible, a little of the adjacent cheek. This will apply direct pressure in 90% of cases, to the bleeding point, and bring about immediate stoppage of bleeding. If the nose is then held for a full 15 minutes by the clock, in most cases, clotting will occur in the vessel, and on releasing, the bleed will have ceased.

Key Points

(1) Pinch the soft part of the nose below the nasal bones; not the nasal bones themselves.

(2) Keep the pressure applied continuously for 15 minutes.

Factors to Reduce Nasal Blood Flow

The instillation of nose drops markedly reduces nasal blood flow, whilst the application of cold to the nose, face or neck as with an ice bag or bunch of keys, does produce some shrinking of the blood vessels. The patient should recline on a bed or sofa, in a semi-recumbent comfortable position, and be encouraged to relax. Nasal packing is only a first-aid measure but cautery by an Ear Nose and Throat Surgeon is a definitive treatment and prevents recurrences.

SINUSITIS

This condition is diagnosed when there is actual pus in one or more of the sinuses. In the majority of cases it occurs in the sinuses of the upper jaw, the maxillary sinus and less commonly in the frontal sinus, above the eye. When the infection is established in the maxillary sinus, it also affects all the other sinuses to a lesser extent as most of these have their drainage openings into the same area of the nose. The infection spreads from the nose into the sinuses and by the time a patient presents for treatment the original nasal infection is often a minor condition or has even disappeared.

Signs and Symptoms

These are nasal discharge, usually coloured and thick, and nasal blockage due to the congestion. This congestion can lead to Eustachian tube blockage. There may be aches or pains around the face, eyes or forehead, worse in the acute phase. Sometimes, there are general symptoms of fever, headaches, tiredness, lack of energy, with aches and pains in the arms and legs. Common signs are pus in the nose and tenderness over the sinuses, while X-rays will show (Fig. 18) the presence of any congestion or pus. Usually the symptoms are much less than in a person complaining of "sinus" or nasal catarrh apart from pain. In some cases the person may feel nothing wrong at all.

Case History

Following a series of dizzy turns due to Alternobaric Vertigo whilst ascending from diving, an out-of-town diver was referred for help. On looking into his nose it appeared that his nose was not fully normal. He had no complaints of nasal discharge, no blockage, no headaches and no face-aches. He did, however, complain of tiredness in his arms and legs. His main fault was Eustachian tube blockage and on ascending he could feel the pressure building up in his left ear until he would become violently dizzy. When he felt the air escape down his Eustachian tube with a hiss, his dizziness would stop immediately. Because infection in the sinuses is a cause of Eustachian tube blockage his sinuses were X-rayed and both were opaque, that is they contained pus. A sinus washout produced pus from both left and right upper jaw sinuses, worse on the left. With treatment he began to feel generally much better and eventually after a major sinus operation he was cured. It was unnecessary operations on normal sinuses in the 1920's that introduced "sinus trouble" into folk medicine.

Treatment

Antibiotics by mouth are necessary for acute cases, and those with pain, but antibiotics have never been much help with sinusitis once it is established or becomes chronic.

Local treatment includes measures to reduce the inflammation and congestion, to promote drainage from the sinuses and to keep the nose clean and clear, so that nasal breathing, with its beneficial effect on the normal functioning of the nose can be carried out.

Nasal decongestants, of which "Otrivine", "Drixine" and "Atamol" are the best, can be given. A spray is likely to be used more efficiently rather than drops, although these latter, if administered correctly are more effective. Nasal soaks for a maximum of three days are recommended for established cases of purulent sinusitis when the diagnosis is not in doubt.

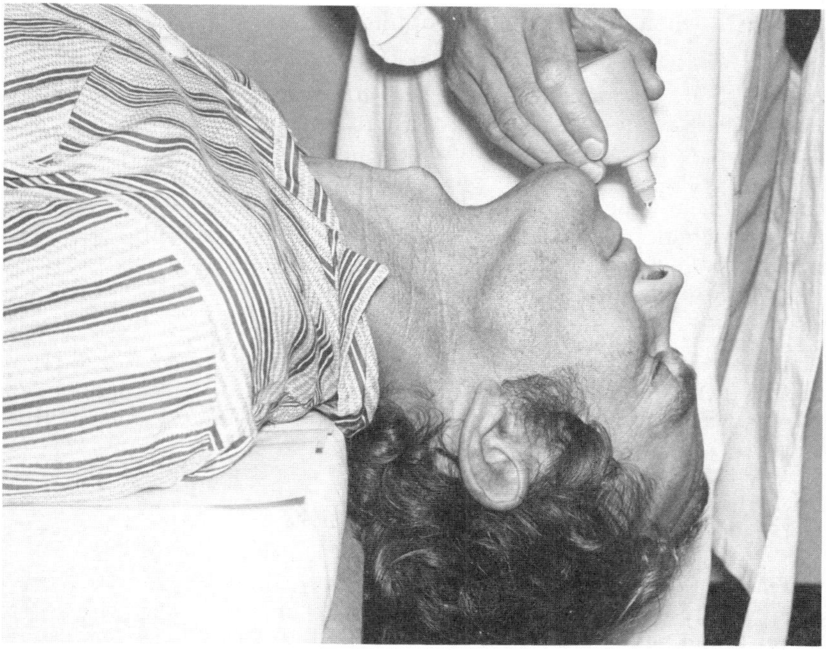

Fig. 19

A sinusitis patient positioned for a nasal soak. The chin is in line with the ear indicating that the head is right upside down. The nasal spray container can act as a dropper.

Nasal soaks are given as for nasal drops but the process is more complicated and must be adhered to very closely if any benefit is to ensue. The patient should lie down on a bed or couch with his head and shoulders over the edge (Fig. 19) so that the head is completely upside down. In this position, put 10 drops in each nostril. These will run directly up into the top of the nose and into the area where the majority of the nasal sinuses open. Keep the head straight back for one minute, twist it to the left for one minute and to the right for one minute. This moves the decongestant drops on to the side wall of the nose further concentrating their action on the sinus openings. If the drops go through to the throat they are wasted as they are meant to stay in the nose. If this occurs then the positioning is wrong.

After a total of 3 minutes by a clock, the patient can then stand up and lean forwards for a few seconds so that the excess drops, if any, will run to the front where they will produce further decongestion and the rest will run out of the nostrils. Nasal inhalations are of some use, especially in acute cases when the warmth is soothing.

Above all, the patient should blow his nose. With an airway that is clear, a fast stream of air passing through the nose is at a lower pressure than the air stationary in the sinuses (Venturi effect). Thus a good nose-blow can actually suck the pus out of the sinuses. It is not the force that counts, not the noise produced, but the speed at which the air is travelling which produces the desired result. If the nose is blocked, prior use of decongestants by tablet or by spray, will open the airways to permit efficient nose blowing.

Usually most cases clear with the minimum of treatment but when the sinusitis persists, as shown by X-ray then specialist treatment which includes washout of the sinuses under a local or a general anaesthetic, is required.

"SINUS SQUEEZE"

One of the other causes of nose bleeds in scuba divers is related to poor pressure equalisation in the sinuses on descent. The walls of the smaller blood vessels, including the capillaries, will not stand up to this pressure differential and they will rupture readily with bleeding into the cavity of the sinuses. On ascent the air expands and forces some of this blood out of the sinuses and it will then come out of the nose.

Thus non-frothy blood coming from the nose or mouth can be coming from the sinuses. This bleeding into the sinuses can sometimes be accompanied by marked pain. An X-ray will show that the sinus is filled with some fluid and the blood would have to be differentiated from pus filling the sinus. This can be done either by examining the nose and considering the way the condition came on or by performing a washout of the sinus which will give an accurate diagnosis.

Prevention is based on having a healthy nose. If there is nasal congestion, then use nasal decongestants, either as drops, spray or tablets by mouth. Do not dive with a cold.

Bleeding does not always occur with "sinus squeeze" especially when the blockage happens on ascent. In these cases there is a build-up of pressure commonly in the frontal sinus with pain developing over one or both eyes. The pain passes off when the air escapes, often with an audible creak or hiss.

Case History

Case J.L. had been down to 80 feet with a bit of a cold which usually had no effect on him. He developed pain in his forehead and because of poor buoyancy control he made a rapid ascent, his mask filling with blood from his right nostril. The pain over his left eye cleared on surfacing, but his head still felt numb. Hospital Emergency Department X-rays showed an opaque right upper jaw sinus (Fig. 18). He was referred to that hospital's ENT Outpatient's Clinic in two weeks. As he continued with left headaches, pussy discharge from his right nostril, and some blockage of his nose he sought earlier help. On the 5th post-dive day he was seen to have pus coming from his right upper jaw maxillary sinus and he was put onto nasal soaks, and a nasal spray. The diagnosis was left frontal sinus squeeze and right maxillary sinusitis plus bleeding into the right sinus. This lattermost was shown up by the fact that after a nasal soak he would blow his nose and pus and black blood would come out of his right nostril. After two weeks he was fully recovered and it is hoped he will be more cautious next time he thinks of diving with a cold.

Key Points

1. A healthy nose leads to safe diving.
2. Do not block the nose when performing nose blowing.
3. Nasal sprays are useful for nasal congestion but only for short periods.
4. Sinusitis can be cured with simple treatment in the early stages.

CHAPTER 7

MOUTH AND THROAT

Disorders of these structures are not common in the scuba diver but some practical points emerge from association and research with underwater swimmers. The teeth and jaws will be dealt with separately.

LIP SORES

The common lip trouble is an ulcer developing on the lip or the gums due to the pressure of a rigid snorkel or breathing tube. As the lip is continuously in salt water, even minor abrasions will take some time to heal.

Causes

These are an ill-fitting or rigid snorkel or the holding of the mouthpiece too firmly. The other common cause is too much exposure to sun and wind and sea. "Cold Sores" or Herpes Labialis can cause severe trouble to the lip in the susceptible person.

Treatment

In the person whose lips are susceptible to exposure, to sun and wind, there are many lip salves commercially available. The best of these should contain Titanium Dioxide, a very white compound as in RVPlus (Paul B Elder Co.), Locorten-Vioform paste (Ciba). These compounds are very good for covering any abrasion or lip sore and should be applied before and after each dive. For ulcers on the inside of the lips or on the gums, Kenalog in Orabase (Squibb), a plastic-based ointment which adheres to wet surfaces is recommended. For "Cold Sores" or Herpes Labialis, Acyclovir Ointment is very effective in curing and healing.

Prevention

This is the best form of treatment and the diver is well advised to be careful to equip himself with gear compatible with his own requirements and this would include a snorkel of non-rigid material that fits into his mouth without pressure on the lips or gums.

If there is a tendency to sore gums from mouthpiece pressure, then the diver must consciously remember not to bite the mouthpiece so tightly.

"Cold Sores" due to Herpes Simplex virus can be a problem and if they occur frequently a skin specialist may be needed for advice. However Acyclovir ointment applied frequently at the first feeling of an itch or tenderness will prevent the blister from developing and reverse the reaction to the Herpes Simplex virus.

CHRONIC CATARRH OF THROAT

All conditions of the throat occur in divers as in the general population, but chronic catarrhal inflammation of the throat is a condition which should be eliminated in divers in order to prevent complications arising from a sensitive throat. After an acute throat infection as accompanies viral infections of the throat and nose, the throat is left in a weakened state. It can also be irritable or feel as though there is a lump in the throat. Because of this irritation the person may develop a cough but more commonly vigorously tries to clear the throat.

This clearing of the throat does more damage to the throat and prevents it from healing up after the viral infection. Thus the irritation persists and the patient does not recover from their irritable throat. Recovery is further delayed by mouth breathing and in some people reflux of gastric juice up the gullet. This reflux is a form of indigestion usually known as heart burn. The acid irritates the throat and prevents healing of the viral infection. People commonly blame the irritation on the "phlegm in the throat" but this is not true. Phlegm or mucus is needed in the throat to lubricate it and keep it moist.

Signs and Symptoms

The patient complains of an irritation in the throat. Some people also have a burning, soreness or raw feeling in the throat. It may be merely a dryness or stiffness in the throat, with no difficulty in swallowing. With chronic inflammation the patient may also have a feeling of something in the throat. The signs vary with the acuteness of the inflammation. There may be a generalised redness, or the redness may be localised to the back of the voice box. Glands in the neck are not often involved.

Treatment

This is often difficult to treat as it requires a change in life-style for the patient. Thus the patient must do what he is asked. He must breathe through the nose and any nasal disorder treated. The simple treatment is the use of a nasal decongestant spray. Reflux of acid gastric juice is investigated. If it is present active antacid treatment is needed and this means more than the usual Quickeeze or Soda Mint which basically are chalk or calcium carbonate. Above all the patient will not get better until they give up clearing their throat. This used to be called Hemming and Hawking. Each time a person clears the throat they beat the two sides together. The lining of the throat is not designed for this and is damaged. This damage is a form of inflammation. With inflammation there is some swelling which is felt as an irritation or a lump in the throat. If there is this irritation it is often stopped by taking a sip of water or by sucking a peppermint or throat lozenge.

ADENOIDS

The persistence of the adenoids (Fig. 20) in adults is rare and does cause some signs and symptoms. These symptoms include mouth-breathing, snoring, noisy breathing at night, headaches, nasal obstruction and nasal discharge, chronic cough and symptoms referable to infected sinuses and, most important of all, symptoms referable to ears. The treatment of resistant sinusitis after failure of nose blowing, breathing exercises and sinus washouts, is the removal of the adenoids, at which time it may be necessary to perform other operations on the sinuses. Difficulty in clearing the ears improves after removal of the adenoids in adults. See Case History on page 30.

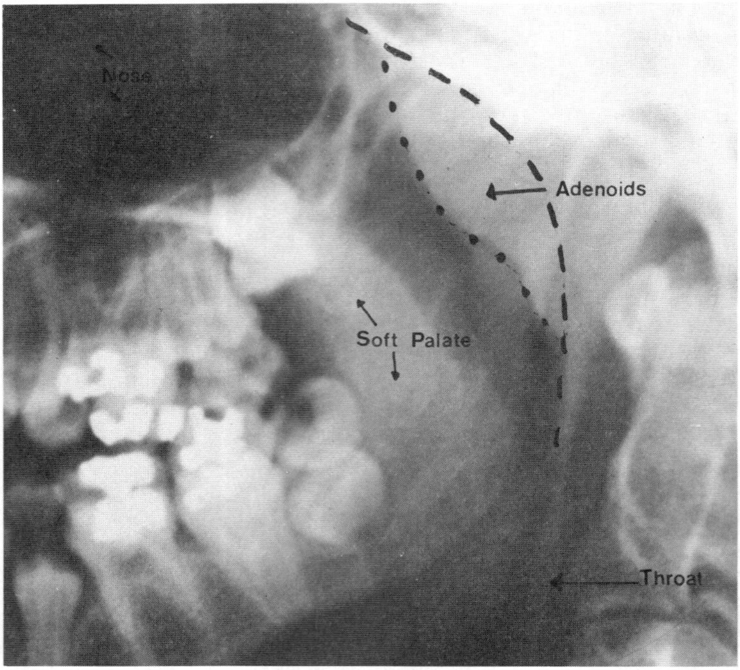

Fig. 20

Adenoids (dotted) as shown by a side on X-ray of a young adult. By taking up half the available space between the soft palate and the dashed line of the nasopharynx they not only block the free flow of air through the nose but also cause congestion in the nose. The last teeth nearest the soft palate are the wisdom teeth. In the X-ray they have no roots and appear to be impacted.

CHAPTER 8

THE JAWS AND HEADACHES

The commonest cause of earache in adults is referred pain to the ears from a disorder of the working of the jaw joint known as the temporomandibular joint. This joint is situated just in front of the ears and joins the lower jaw or mandible to the upper jaw (Fig. 21). The back wall of the joint is the same as the front wall of the outer ear canal. Put a finger in the ear and open your mouth. You will feel the front wall of the ear move. The upper joint surface is on the underside of the temporal bone of the skull which contains the middle ear and the inner ear. Because these structures are all close together, their nerve supply comes from the same nerve branches. For example, the one branch of the Fifth cranial nerve which supplies the three main jaw muscles, also supplies the eardrum muscle of the middle ear. The nerve of sensation of pain to the jaw joint is the same nerve which provides sensation to the outer ear canal and the eardrum. A referred pain is one arising from one particular place, but is referred or placed by the brain to another place. For example, the pain from an impacted wisdom tooth is usually felt deep in the ear.

JAW JOINT DISORDER

This was formerly known as Costen's Syndrome but its cause and distribution of symptoms has gone far beyond that envisaged by James B Costen of Harvard in 1933 and subsequently. This disorder is poorly understood because its causes lie in a region which falls between the jurisdiction of the doctor and the dentist, neither of whom are taught much about its consequences. Also there is nothing in folk medicine about the jaw joint so most people have trouble in comprehending the problem; even as do many doctors. Scuba divers are very conscious of aches and pains in and around their ears and faces because of the trouble the ears and the sinuses give them, so they report for treatment for such disorders more readily than the average patient. Earache may occur unexpectedly at times in a "normal" ear which has become sensitive due to other factors. At least one third of the divers seen at my office have this problem. In some it is their only problem and in other it may only be a minor part of their trouble.

Case History

This is what one lady wrote to me. "When you told me that it was my teeth and jaw causing trouble I thought that it was just a lot of hot air. Since I have been doing your exercises I have had no further trouble". Another patient rang up to report her cure but said "Miraculously my pain (earache) has gone". She credited the cure to a "miracle" rather than my treatment.

Causes

One of the causes in underwater swimmers is forceful clenching of their teeth on the mouthpiece of their snorkel or their regulator. Some divers even bite the nipples right off. This may be an unconscious effort to retain a hold on an essential-for-life piece of equipment, in their uniquely stressful environment, or it may already be their habit to clench their teeth.

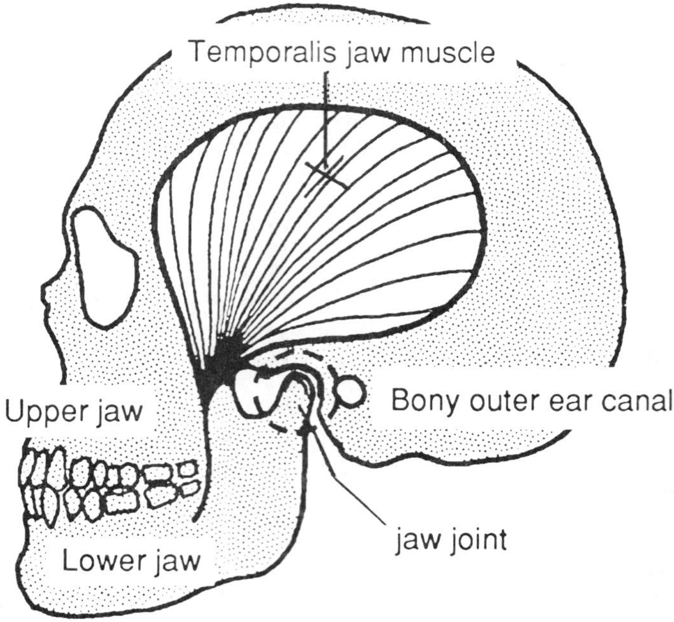

Fig. 21

The main lower jaw muscle - Temporalis. It extends from the temples in front, a common site for pain, to the skull behind and above the ears. The back wall of the jaw joint is the front wall of the outer ear canal.

The onset of symptoms is probably due to the presence of several factors working together. Thus it is likely that there is a tendency in affected divers to grind or clench their teeth even when out of the water. If there is faulty dentition due to lack of teeth or their mal-position, or if the dentures are unsatisfactory, then there is an abnormal state existing. This abnormal state is known as malocclusion. This then causes subconscious irritation leading to faulty habits of moving the jaws. Rarely is the joint physically affected but the many muscles which move the jaw joint tend to develop movements to protect it. These muscles go tense or tight and put pressure on the joint. If they go too tight or into muscle spasm then pain arises from the muscles similar to back pains and lumbago. Pain then will arise at the site where the muscles are attached to the bone of the skull, the jaw, or in the muscle itself. The cold of immersion into the sea always aggravates muscle pain which is relieved by warmth. Muscle tension from anxiety in the beginner is another factor.

Symptoms

Pain, varying in intensity from dull aches to sharp stabbing pains occur anywhere over the side of the head, in or around the ear or down the neck, in the throat or along the side of the tongue. An amazing possible distribution of pain. Pain or aches in the temples above or below the eyes also come from the jaw muscles. The pain over the side of the head is related to the extensive fan-like insertion of the temporalis muscle (Fig. 21) one part of which only may be affected. It may also occur in individual teeth due to a non-infective inflammation of the pulp brought about by tapping of one tooth in the lower jaw by one tooth in the upper jaw. This is not uncommonly done by people in their sleep. This makes the tooth sensitive to cold in which case the sudden inhalation of cold re-expanded air may set the pain going.

This pain is made worse by cold and will often occur at the beginning of a dive. The pain or aches may also persist for hours or days after a dive. The patient may complain of deafness but it is usually a blocked feeling in the ear rather than any loss of hearing. Dizziness, more of a non-specific lightheadedness or "muzziness" is often mentioned by the patient. The dizziness is partly the physical sensation felt when the muscles of the scalp are held in a state of tension or abnormal contraction. The mental tension or stress of diving also causes these muscles to tighten and aggravate the light-headed feeling.

Signs

These are the lack of molar teeth to support the jaws, impacting wisdom teeth or ill-fitting dentures. Impacted wisdom teeth or the third molar teeth, are unable to erupt because of lack of room in the jaw. Excess wear of the front teeth can sometimes be observed and this can only be caused by grinding the upper and lower teeth together, usually during sleep. Tender sites in the jaw muscles can be found such as in the temporalis jaw muscle on the side of the head. Some people have clicking in their jaw joint.

The eardrum, despite what may be intense earache, appears normal, unless there is another condition present at the same time. Signs of barotrauma or the squeeze of the eardrum are commonly present if the diver comes soon after a dive. Although barotrauma and pain go together it is only when diving that pain is caused by pressure. Once the pressure is normal then the pain may not persist even though the eardrum does not appear normal.

Treatment

An explanation of the cause with cessation of clenching, grinding or one-sided chewing often results in relief. In any case with obvious dental disorders, treatment should be sought from a dental specialist. Jaw muscle stretching exercises, by taking the tension out of the muscles are very helpful (see page 91).

Case History

A mixed group of professionals including several doctors were having a mid-week dive at the Poor Knights. One pair swam about 100 yards over to the entrance to the big arch before going under but soon they resurfaced because one of the pair developed severe pain in his left lower fifth (pre-molar) tooth made worse with each breath from his two-stage reducing valve regulator. He had been puffing a bit as the result of the surface swim and because of his continuing deep breathing under water he gave his tooth a good dose of the cold air. He could not resurface immediately but had to catch up with his buddy and let him know he was returning to the surface. With a five minute rest and a slow descent the pain did not come back.

When he returned to the boat after the second dive he mentioned to one of the other doctors his episode and also the fact that his teeth did not seem to fit together as they should. This doctor said that his never did after one hour of diving, and that often he developed an ache up the back of his jaw and into his ear.

The first diver had been grinding his teeth, so his wife later told him, and this had made this one particular tooth sensitive to the cold. The doctor's remarks were further proof of how, by keeping the jaws held firmly in an unnatural position, that is clenching the mouthpiece, the muscles tried to adapt. Excess adaptation caused the muscle pain, whilst the whole balance of the jaws was changed after two hours of diving so that there was difficulty in getting the teeth to meet naturally again for an hour or so after the dive.

SCUBA HEADACHES

Headaches can be a very real problem with many divers and some types can be attributed directly to the scuba diving, to the gear or to the actual dive. An over-tight mask strap by continual pressure on the head, can cause a headache. Swimming long distances with the head tilted backwards can cause headaches. In the past the contamination of the air by poor air compressors, caused many headaches. This would be due to contamination with carbon monoxide from a faulty or poorly placed exhaust of a petrol driven compressor. The majority of compressors are oil lubricated and unless the oil is well filtered and frequently changed, overheating results in the production of the carbonic gases from the oil and these contaminate the air. Both types of gases are poisonous.

In carbon monoxide poisoning, the early symptoms are difficulty with breathing when physical effort is made, tiredness and lack of energy, with headache, nausea, giddiness and palpitations of the heart. Treatment consists of providing fresh air and oxygen, preferably hyperbaric oxygen for severe cases. Prevention is more important and this means using approved air filler stations and taking a few breaths to test the air cylinder before entering the water. Carbon monoxide, especially from exhaust fumes, does have a distinctive flavour.

The build up of carbon dioxide in the body does occur with over-exertion at depths, more so with the use of skip-breathing techniques. This build-up is due to poor lung ventilation, and produces flushing of the face and headaches. To overcome this, a deliberate increase in the breathing rate should be made when physical activity is increased at depths.

Case History

Diver D. W. when uplifting gold bullion from the Eligamite noticed this flushing with headaches. He realised what the problem was and by breathing more frequently his troubles were relieved. When performing physical work at depths he now purposely over-breathes and does not use the skip-breathing technique. Skip-breathing is okay for amateurs slowly meandering along.

Changes in blood pressure, by upsetting the intra-cranial circulation inside the skull, can result in headaches, the pain of which probably arises from pain endings in the walls of the blood vessels. Lying down and especially lying in the hot sun, lowers a normal blood pressure, and in the susceptible person, sudden activity or immersion into cold water can then cause a sudden blood pressure change with a resulting headache.

One of the side effects of scuba diving in some people is the increased urinary excretion with an actual dehydration effect on the body. This dehydration can also lead to headaches so an adequate non-alcoholic fluid intake is advised. Alcohol does affect the kidneys and part of the "hangover headache" is due to the accompanying loss of fluid and salts from the body leading to dehydration.

Consequently, when a scuba diver develops headaches it is likely that there are several causes acting at the same time and it is up to the individual to eliminate some of them by careful preparation and avoiding those factors mentioned in this section.

SINUS HEADACHES AND FACEACHES

Some divers have smaller openings than usual into their sinuses (see Fig. 17). If these block off then pressure changes do not occur in the sinuses and sinus squeeze occurs. This happens on descent or on ascent. This stimulates the nerve endings causing pressure feelings, aches or pains which are felt over the face, in or around the eyes, in the forehead, temples or on top of the head. Thus quite acute headaches do arise from completely uninfected sinuses. Headaches from purulent sinusitis are uncommon and mild. The headaches said to be due to "sinus" or "sinus trouble" are, in the non-diver, due to the jaw joint disorder or a typical migraine. Air trapped in a filling in a tooth can result in toothache or faceache with pressure changes.

MIGRAINOUS HEADACHES

True migraine is well-known even to the patient, but questioning to see that it fits the classical pattern is always necessary before accepting the sufferer's diagnosis. A person who suffers migraine should not take up diving. Its onset is bound up with carbon dioxide levels in the blood and an attack can be precipitated by overbreathing or hyperventilation. This overbreathing can be set off by immersion in cold water or by rapid breathing due to fear of the watery environment. Also the use of a snorkel means that some of the expired air in the snorkel is breathed back in. This air has a high level of carbon dioxide, thus increasing the blood carbon dioxide level and bringing on a migraine attack.

Diagnosis of Headaches

Headaches are a common complaint of scuba divers and many of their causes have been mentioned. Because it is common practice to try and fit a particular patient into one or another of the two common varieties of headaches, it would be of advantage to know what the common varieties are. The diagnosis of the headache is almost totally dependent upon the ability of the doctor in extracting a detailed history of the diver's complaints and to recognize specific patterns of symptoms. To the neurologist, the tension headaches and vascular headaches are the commonest. To the physician neuralgias, causalgias with burning sensations and migraine would be the commonest. To the ear nose and throat surgeon, the commonest cause of headaches is from the disorder of the muscles controlling the jaw joint. The distribution of the muscle attachment of those muscles moving the lower jaw is widespread over the side of the head and all degrees of pain from severe sharp jabbing pains to dull aches can arise from the jaw muscles either localised or widespread. Thus a disordered set of teeth with excess clenching or grinding of teeth, especially in the presence of emotional tension, can be responsible for a wide variety of headaches and pains. Less common are the headaches from the sinuses. To some people a hang-over headache could be the commonest. Eye strain, high blood pressure and neck problems are rare causes of headache.

In general a headache should be fitted into one of the common groups and then treated accordingly. If a headache does not fit into one of these groups or if the response to treatment is not satisfactory, further investigation may be indicated.

Key Points

1. There is nothing in folk medicine about the jaw joint so divers have trouble in understanding this problem.
2. Do not bite hard on the mouth piece of the regulator.
3. Common causes of headaches can be easily avoided.
4. Headaches and faceaches from "normal" sinuses are not uncommon with pressure changes.

CHAPTER 9

HOW TO CLEAR THE EARS
or Prevention of Middle Ear Barotrauma (The Squeeze)

There are approximately 120,000 sport scuba divers certified and active in New Zealand. The comparable figure for the United States is 1.2 to 1.4 million. The commonest problem with divers is barotrauma or pressure damage to the air containing spaces of the head and is experienced to some degree by almost all. Pressure damage or barotrauma works in various ways. On descent the water pressure pushes the eardrum inwards and this eventually becomes painful. That is middle ear barotrauma or the "squeeze." Further increase in pressure can cause the eardrum to burst inwards and that is severe barotrauma or damage. The ear is the most common part of the body to be affected by barotrauma. In researching this topic the only informative article was that of Delonca (1970), "Considerations sur les manouvres dites d'equilibration de l'oreille chez le plonguer" and this was written in French.

Most authors mention Valsalva Manoeuvre but do not specify what they mean by this. In his treatise "Tractus de Aure Humana," published in 1704, Antonio Valsalva described a procedure which consisted of increasing the pressure in the chest by forced- breathing out with the mouth and nose firmly closed, thus causing the air to pass into the middle ear by way of the Eustachian tube. This procedure was originally used by Valsalva to blow pus from the middle ear out through the perforated eardrum. This procedure with adaptations is still very effective in the treatment of discharging ears and otitis media with effusion or glue ear.

In view of the lack of written information on this topic, yet its considerable importance to scuba divers and their instructors, this detailed chapter has been written. For the many non-diving doctors this information would be of extreme importance when dealing with divers having problems with "clearing the ears" or forcing air up the Eustachian tube to equalise middle ear pressure with surrounding pressure on descending underwater.

How I Went About It

In my private practice, over 1,200 scuba divers have been seen for various problems and 60% were treated for middle ear barotrauma. These cases either had visible evidence of ear damage or their story contained the fact that they suffered pain in either one or both ears when they were underwater. Active personal diving over a 30 year period with continually probing and researching the mechanics of the Eustachian tube has provided me with further information.

The first step is to determine whether the pain was a primary ear problem due to Eustachian tube problems, or was secondary to other problems (see Chapter 8). If it is certain that the pain is due to the Eustachian tube problem, it is necessary to separate out the cases into three groups.

Group 1. The squeeze is only due to physical disorders such as primary Eustachian tube defect, or secondarily due to nasal congestion, unilateral or bilateral, or a crooked nasal septum (Fig. 16). Rare physical causes are persisting adenoids, repaired cleft palates and nasal polypi, these latter not always causing any disturbance to Eustachian tube function (55%).

Group 2. The barotrauma is only due to a disorder of the working of the Eustachian tube. This group includes those whose technique of "clearing the ears" or forcing air up the Eustachian tube, fails to work and there are no physical disorders present (23%).

Group 3. A mixture of both Groups 1 & 2, that is physical disorders and poor technique (22%).

Table 2
Preventive treatment of 399 middle ear disorders

Cautery of the Inferior Turbinates of the Nose	
As the Only Treatment	200
Plus counselling on technique of clearing the ears	87
	—
	287
Counselling on the technique of clearing the ears only	57
Miscellaneous treatments	
Deep x-ray therapy, drugs, adenoidectomy, submucous resection of the nasal septum, septoplasty, nasal polypectomy	20
General advice on safer diving. eg., not with a cold	35
	—
TOTAL	399

Table 2 is an analysis of the treatment of 650 divers. This gives the distribution of these groups and includes a group "General advice on safer diving." This group were given counselling on technique but were mostly those who had dived when they had a common cold or other nasal infection. Other tips for improving the ability to clear the ears are to refrain from lying flat between dives as in sunbathing. Lying flat causes nasal congestion, as also may the drinking of alcoholic beverages between dives, by some people. Nasal congestion blocks the Eustachian tube. Thus 55% were in Group 1, 23% in Group 2 and 22% in Group 3.

Classification into Groups

Classification is carried out in a 3-step examination. **Firstly** the patient is asked to perform the Toynbee Manoeuvre. This was described by Joseph Toynbee (1853) as a method of determining Eustachian tube function. With the nostrils pinched closed, the diver is asked to swallow. If the diver feels a pressure in the ear or hears a click then the result is declared positive, and normal Eustachian tube function for that ear is accepted provisionally. **The second step** is to ask the diver what is the exact technique, moment by moment that he uses to "clear the ears." Some divers have difficulty describing this, and patience is required especially not to make any positive suggestions to which they often readily agree. Their technique should follow what will be described later, but the main fault lies in the diver trying to do Valsalva manoeuvre without exerting maximum pressure. In other words they do not blow as hard as they can down the nose. A second rarer fault is those whose Eustachian tubes are always open or nearly so. These divers "clear their ears" so easily that they can be unaware that it has happened. They do not get the usual hiss, click, pop or bang in the ear that they have been told to expect. **The third step** is to carry out a full ear, nose, throat and back of nose examination, sometimes with sinus x-rays to exclude infection therein. This will then pick up those physical disorders which cause poor Eustachian tube function.

Having eliminated other than ear causes of ear pain (see Chapter 8) these three steps then allow classification of the diver and point to the treatment needed. With physical causes eliminated, the technique on how to "clear the ears" needs teaching and this is called counselling. During this counselling the diver does carry out the simpler methods as they are explained to him, and a written hand-out with the main points included, is given to the diver for further study and practice at home.

Counselling

With divers having difficulty in clearing the ears it is a relative problem of the action of the Eustachian tube which makes it difficult to "clear the ears." It is a relative problem as only one of these divers suffered from ear problems when not diving, indicating that the Eustachian tube is adequate for life on earth. This one case had "glue ear" and refused treatment because with a middle ear full of mucus he had no need to "clear the ears." There was no air in his middle ear that would be compressed. He went diving every weekend with no problems. He was much like the seal, whose middle ear has large veins in the lining, and these fill with blood as the air volume decreases.

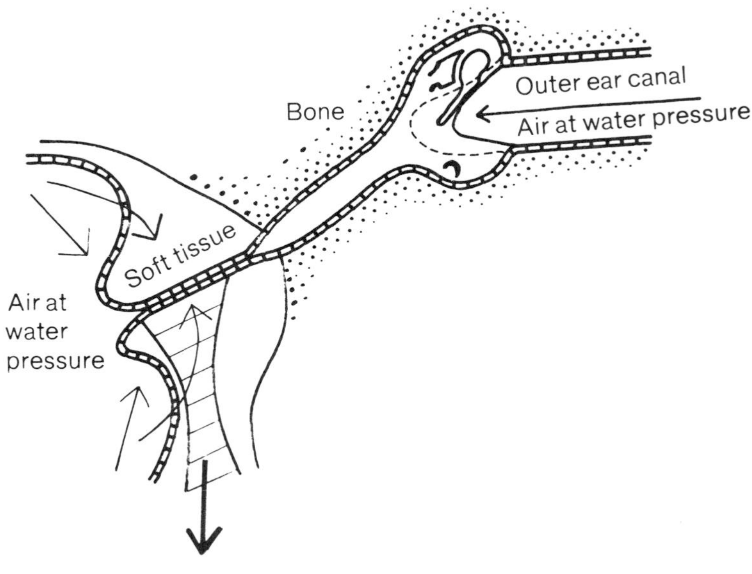

Fig. 22

The Eustachian tube locked closed because the water pressure is greater than the muscle opening power of the soft palate muscle (dark arrow).

VALSALVA MANOEUVRE

Counselling consisted in training the diver in Valsalva Manoeuvre, then moving on to Frenzel Manoeuvre and even the French "Voluntary Tubal Opening," all of which are mentioned in this chapter. Two other methods will also be described.

To do **Valsalva Manoeuvre** the diver pinches his nostrils closed, although if he does not have a mask shaped to his nose, it may be enough to press the rubber of the mask hard against his nostrils to block them. Then the diver blows hard down his nose against this blockage. This is successful in forcing air up the Eustachian tube in some divers, but the success rate is further increased if the diver swallows at the same time as he is blowing. However to swallow whilst the teeth are propped apart by the studs of the regulator held between the teeth, is difficult and in most divers impossible. In proof of this try to swallow with the mouth open.

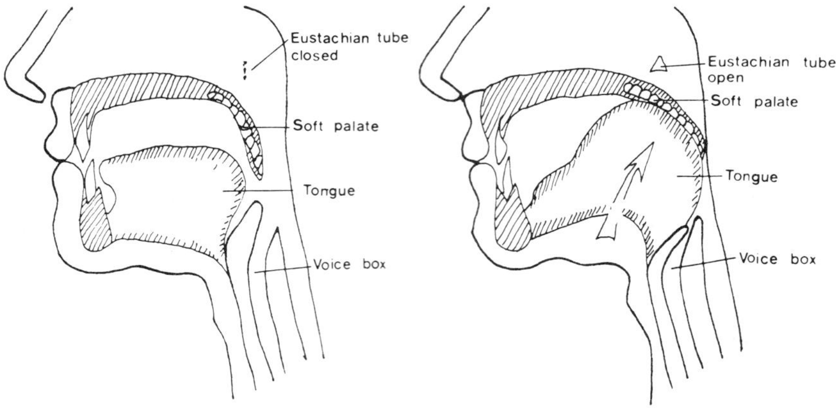

Fig. 23

Frenzel Manoeuvre. Left - tongue and soft palate at rest. Right - tongue pushed back and upwards with the nose blocked.

That Valsalva Manoeuvre succeeds in many divers is against the biomechanics of the Eustachian tube. From the diagram (Fig. 22) it is clearly seen that as the water pressure rises with descent so does the air pressure in the throat at the back of the nose. This pressure is the same as the inspired air pressure coming from the regulator which automatically delivers air from the air tanks at the surrounding water pressure, that is atmospheric pressure plus the pressure of the column of water from the surface down to the diver.

The inner inch (2cms) of the Eustachian tube is surrounded by soft compressible tissue. As the pressure rises on this it squeezes it closed. If at the same time the diver blocks his nostrils and raises the pressure this further clamps the tube closed. It is estimated that at a depth of six feet (2m), the water pressure closing the tube is greater than the pressure that the expiratory muscles of the chest can produce, so the Eustachian tube is then said to be "locked." Valsalva Manoeuvre will be successful if the inner end of the Eustachian tube is, under these circumstances, opened by an increase in air pressure above that of a forced expiration or by muscular action or by both. Further manoeuvres to do this are now described.

FRENZEL MANOEUVRE

This problem was studied by Dr. Herman Frenzel of the Luftwaffe in 1938 because of the ear problems dive-bomber pilots were having with their high-speed descents. His method was to train the pilot to close his vocal cords (Fig. 23) and to compress the air in the back of the nose by forcing the back of the tongue upwards and backwards like a piston, pressing the soft palate up against the sides of the throat. To perform this, one takes a light breath in and then close the vocal cords together suddenly as if about to lift a heavy weight. Pinch the nostrils closed and start to make the sound "K." This then raises the base or back third of the tongue, forcing it up against the soft palate which in turn compresses the air in the back of the nose and in the nose. This pressure can be enough to force open the Eustachian tube and allow air into the middle ear. The mouth may be open or shut. This manoeuvre is completely independent of respiration or breathing and has no effect on the air in the lungs below the vocal cords. There is also no effect on the flow of blood into the chest from the neck, thus preventing any increase inside the skull of the cerebro-spinal fluid pressure. A rise in this pressure can cause rupture of the Round and Oval Windows.

VOLUNTARY TUBAL OPENING

The third method relies upon the activation of the muscles used in the opening of the Eustachian tube which is normally closed or blocked. Mention has already been made of swallowing, during Valsalva Manoeuvre, and this normally produces opening of the Eustachian tube. French naval diving schools have taken this a step further in the method known as "Voluntary Tubal Opening." By contracting and holding the contraction of the soft palate muscles and the upper throat muscles, a muscular force capable of holding the Eustachian tube open, is obtained. This allows automatic equalisation of the air pressure between the back of the nose and the middle ear. This manoeuvre is said to be possible for the trained diver with a favourably formed and shaped Eustachian tube.

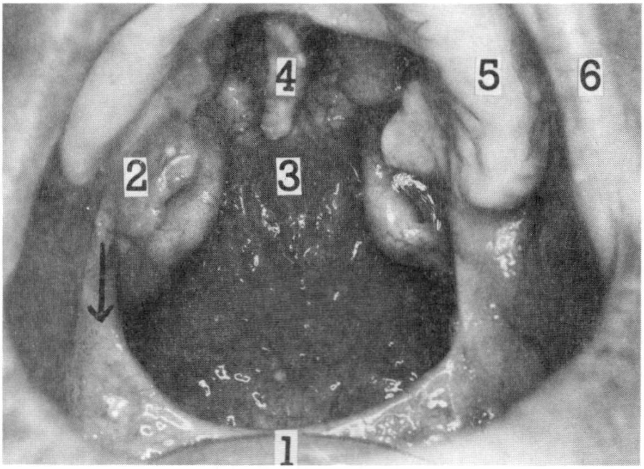

Fig. 24

Photograph of the inner ear end of the Eustachian tube (2) taken through the mouth in a patient who has had most of the roof of his mouth (the hard palate) removed. Note the Eustachian tube opens in the middle of a bulge known as the Eustachian cushion which can be viewed as in Fig. 2. The arrow indicates the muscle passing from the front of the Eustachian tube down into the soft palate. When the soft palate is pulled down or is lifted up as in swallowing, this muscle pulls open the Eustachian tube. The tongue (1) is against the remaining soft palate. (3) is the region of the nasopharynx or the top of the throat, not usually seen as it is above the soft palate, at the back of the nose. (4) is the nasal septum which divides the nose from front to back into the left and right sides. Where the septum ends is the back of the nose where it opens into the top of the throat. (5) is the gum of the upper jaw and (6) is the upper lip. The special upper denture worn by this patient completely seals off the nose from the mouth.

The technique requires much practice. One begins by taking notice of the feeling of the soft palate muscles and the throat muscles at the moment of sensation of fullness in the ear that is to say, during the opening of the Eustachian tube as performed by any of the other methods or during a yawn. Once this feeling is noted the diver practices until it can be produced on demand. Thus the tube can be held open. Needless to say this is a difficult method to teach, impossible for at least 50% of divers but obviously the best method.

The fourth method is a personal one and is easier to teach. The muscles opening the Eustachian tube are in the main, the lifter of the palate (Fig. 24), (Levator Palati) and the tensing muscle of the palate (Tensor Palati). These contract when one starts a yawn or a smoker blows smoke rings. Having then discovered this feeling, one can voluntarily start a yawn. To help the person learn this I show them my soft palate with the uvula hanging down in the resting state (Fig. 25A). Then I contract my soft palate muscles hearing a click in my ears as the Eustachian tubes open momentarily. At this stage the uvula is tilting forwards (Fig. 25B). The normal curved arches or the folds passing from the uvula and soft palate to the side of the tongue, flatten out and descend slightly. Once the diver sees this movement clearly he then places his finger under my jaw and I repeat the movement. He feels the contraction of the muscles in the floor of the mouth being the muscles of the tongue.

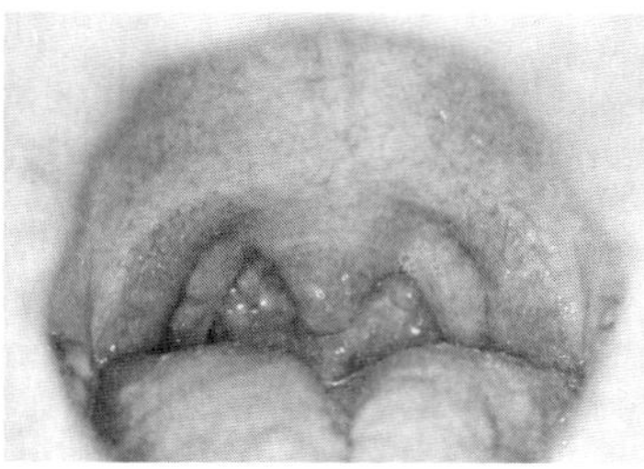

Fig. 25A Veiw of the tongue, the throat and the soft palate with the central uvula. On either side are the curved front and back folds from the soft palate to the tongue. All are in the relaxed state

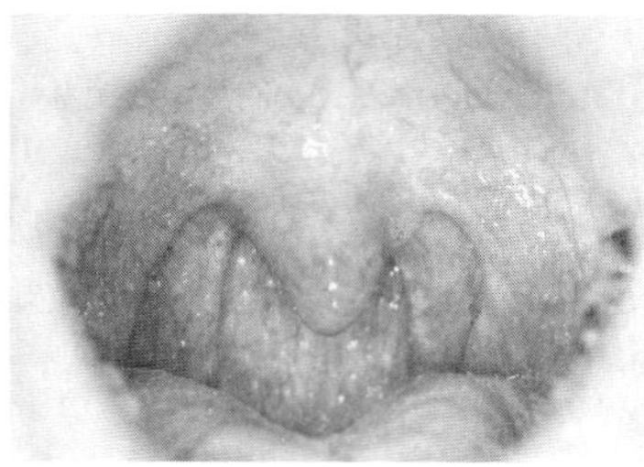

Fig. 25B The same view after contracting the muscles which open the Eustachian tube. Note that the tongue has descended and this can be felt by the diver putting the tips of his fingers under the jaw. The curve of the folds has decreased and the back fold is almost vertical. The width of the throat is reduced. The uvula has tilted forwards.

The diver is then asked to practise at home by looking in a hand mirror with his back to a window. The light is reflected off the mirror into the mouth showing up clearly the palate, the uvula and throat. The diver then tries to mimic the movements. Once he succeeds he can add a thrust forward of his lower jaw. In effect the jaw pulls on the tongue muscles, which pull on the soft palate muscles, which pull open the Eustachian tube. This is similar to the "Voluntary Tubal Opening" but experience has shown that it is the rare diver who can maintain the tube open. Most divers can do it for a split second.

To carry the Roydhouse Manoeuvre out, the nose is blocked and the diver blows down the nose as he does this movement of the palate. The pressure plus the momentary opening of the Eustachian tube allows air to go up and the pressure is equalized.

Whether or not a person succeeds in holding the Eustachian tube open, he should at least learn to activate the muscles which tend to open up the tube so an increase in pressure in the nose or throat finishes off the opening. This then enables "clearing of the ears" to be performed without blowing so hard against a blocked nose.

THE TWITCH

The fifth method is known as "The Twitch," and is the easiest to teach. With the nostrils blocked and a moderate rise of pressure in the nose, the head is quickly and suddenly twitched sideways. To the left to clear the right ear and vice versa. It has to be a sudden, quick, firm sideways movement of the head. It works better if the muscles at the back of the mouth are tightened as in the fourth method. It is especially recommended when there is difficulty with one ear. Failure of this to work is caused by the diver turning the head during the movement or too slow an action.

Discussion

As far as physical or anatomical disorders discovered are concerned, not all defects need correcting to ensure Eustachian tube function. Even after a thorough and radical straightening of the crooked nasal septum some divers still have difficulty in "clearing their ears." These divers need teaching of techniques of "how to clear the ears" as operations are not always successful. One of my divers with complete nasal obstruction due to nasal polypi had no difficulty in "clearing his ears." Even when some divers perform a positive Toynbee Manoeuvre, treatment is indicated if the physical disorder is marked.

Valsalva Manoeuvre can be dangerous if the forced expiration is maximum and held on too long. There are multiple effects on the dynamics of blood flow and cerebro-spinal fluid pressures inside the head. Blood pressure changes can cause fainting. The diver is warned not to hold his forced expiration any longer than five seconds, after which time he should take a breath and have another try. If the fluid pressure inside the head is raised for too long, then rupture of the Round and Oval Windows can happen (Chapter 5).

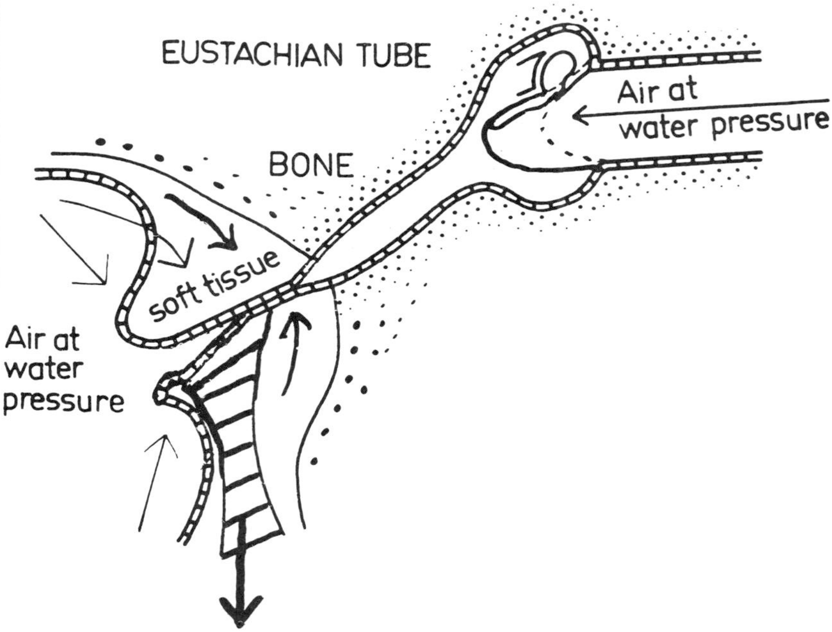

Fig. 26

The Eustachian tube half open. The soft palate muscle (bold arrow) has been contracted or activated so that the inner end of the Eustachian tube is partly opened.

There are other items to counsel the diver on. He is advised to "clear the ears" on the surface and at every 2-3 feet (1m) on descent. If there is difficulty at any depth the diver is advised to ascend to a level at which "clearing the ears" can be performed. If the divers have a cold in the nose or extra nasal congestion from an attack of nasal catarrh (vasomotor rhinitis) or excess alcohol intake and are unwise enough to dive, they should not persist diving if there is any difficulty in "clearing the ears."

Conclusion

"Clearing the ears" can be a problem to most scuba divers especially beginners. Those with the best nose and Eustachian tube function may have only to breathe down through the nose into the face mask to provide enough back pressure to aerate the middle ears. Others will find Valsalva Manoeuvre sufficient, aided by swallowing.

If this is not sufficient then the diver should learn controlled activation of the muscles of the soft palate and tongue by observing these structures in a mirror and getting to know the feel of their muscles in contraction. A combination of this with Valsalva's increased thoracic pressure, can work. Frenzel Manoeuvre may be needed. For those who wish further ease in "clearing the ears," Voluntary Tubal Opening can be attempted. "The Twitch" is a simple method and if it is successful safe diving is ensured.

But above all practice in and out of the water, with or without a regulator mouthpiece in position propping the jaws apart, will lead eventually to quick and ready ability to "clear the ears." This will then prevent the common scuba diving problem of middle ear barotrauma, or the middle ear "squeeze."

CHAPTER 10

VERTIGO, DIZZINESS AND NAUSEA

Vertigo has been reported in underwater divers for decades but Lundgren in 1965 named it as Alternobaric Vertigo. Since 1965 it continues to be reported and various theories as to the causation have been described. Lundgren found this disorder to be common in divers both recreational and professional. He offered no real cause for it apart from it happening on ascent in those who had some blockage of a Eustachian tube. Divers can usually pick out the particular ear that is causing the problem. The prime cause is a change in pressure or alternating (Alterno) pressure (baric). In scuba diving the pressures can change rapidly. This vertigo does also occur in fast descents in flying personnel. Most divers refer to this as dizziness when they describe what has happened to them. Both vertigo and dizziness are symptoms and they are not a unique disorder as they can both be caused by many different disorders (see index & glossary). These words refer to some form of unusual sensation that the diver experienced. Because the diver is upset by these sensations he seeks help.

Case No. 380 was a 28 year old male with six years diving experience. He ascended from 45 feet (14m) and felt pain and pressure in his left ear which began on the ascent and continued to a lesser extent after the dive. More disturbing to him however was the vertigo he developed on this ascent so that he came in for advice the day after the dive.

Both the words **vertigo** and **dizziness** mean the same according to the Concise Oxford Dictionary, but medically we tend to give them different meanings in order to sort out the cause of the problem. To sort these two out requires questioning of all the little details of which most patients are unaware. Sometimes it is like getting blood out of a stone to get all of these little details. It is not the fault of the divers because they are not medically trained. They like to tell you what they think you should know and this is not always of use in diagnosing. As far as **dizziness** is concerned it is not related to any serious condition and it will be defined as some non-specific feeling. It could be called light-headedness, wooziness or some vague feeling of imbalance or feeling of something not being quite right in the head, or of faintness, and usually difficult to describe.

Case No. 649. This 19 year old lady was doing a diving course. She found that after the swimming in the pool she was dizzy. By this she meant that she saw spots in front of her eyes, felt a little sick and light-headed. She also said that she suffered "vertigo" when she woke in the middle of the night. She would sit up and feel as though she was swaying back and forwards. This had been happening for two years. This is a case of dizziness and certainly not medical vertigo with spinning. Believe it or not, her problems all arose from her impacted wisdom teeth. After their removal she had no further dizziness.

Case No. 36 was a 29 year old harbour board diver who had been diving daily for three years. For the first dive of the day he would have a "squelchy" sound in his left ear on ascent. After one or two ascents the left ear would block on ascent and he would go "giddy" (his word). He claimed that there was no rotation. (Eleven years later he was seen with a recurrence of his problem and he did have a spinning vertigo). The giddiness lasted after he surfaced for up to five minutes. If he stood up he would fall forwards. He would feel "really seasick" with the nausea. This had been occurring for two years and was getting worse. Originally he would only feel "woozy" but now he was "really giddy." He was not a typical case of alternobaric vertigo.

The interesting feature of this diver, and this feature has been repeated with others, is his description of the onset of the giddiness. "I would be coming up the shot-line and when I would be about six feet (2m) from the surface after 30 foot (9m) dive, the pressure would come on in my left ear. The giddiness would then come on. If I went down six inches the pressure would go and the giddiness would stop. If I went up the six inches the pressure and giddiness would come on, so down I would go again. After awhile I would get fed up and would rush up the shot-line, climb on board and lie on the deck. The giddiness would stop with a squelchy sound in the back of the nose as the pressure went from the left ear. The pressure sometimes went on surfacing but other times it would last up to five minutes after surfacing."

Vertigo, on the other hand, has nothing vague about it. The diver usually talks quite emphatically about the way everything was spinning around. He may have had a loss of orientation or direction. Sometimes the diver cannot decide which way is up, noting that "up" is a very nice place to be under such prevailing circumstances. Vertigo is defined medically as that feeling in which there is rotation or spinning in the head or of the surroundings and many divers can indicate which way round the surroundings were going. When it is severe there may be complete loss of orientation and panic and as one diver said, "If it wasn't for the shot rope I would have drowned," even though he was only in 20 feet (6m) of water. I have interviewed many divers who have had this extreme form of vertigo and they had all survived. The loss of sense of direction and disorientation can be a severe psychological upset.

Case No. 544. This diver was seen because of pain developing in his left ear at about 15 feet (5m) and even after equalizing he was left with an ache. He was a diver with 15 years experience and had recently been having trouble clearing his left ear. However he confused me by stating that on a recent occasion he had his pain in the left ear on descent and suffered vertigo. He did equalize and his ear felt odd, bubbled a bit, the pain persisted but the vertigo ceased on further descent. His left ear felt blocked for 3-4 days. He had also had vertigo on ascent with some dives and sometimes nausea.

The diagnosis was left middle ear barotrauma and this was treated by cautery of the left inferior turbinate for the nasal catarrh (vasomotor rhinitis). Owing to the relative Eustachian tube blockage he developed alternobaric vertigo on ascent. In one episode he had a painful left ear on descent and vertigo occurred but it went away on further descent. This is the rare case of vertigo on descent. As there was pain only on the left he held his nose and blew very hard to try and inflate the left ear. However he over-inflated his right ear. This pushed the stirrup (stapes) bone into the inner ear. This then resulted in vertigo. With further descent as the right middle ear equalised the vertigo ceased. Thus the vertigo was due to overpressure in the right middle ear as usually occurs on ascent. The right ear was overpressurized by continuous attempts to pressurize the left ear.

However to complicate matters there can be a grading from dizziness to vertigo. Case No. 36 would only say he was "giddy" when he was first examined. He denied any spinning sensation. However 11 years later he did have the spinning vertigo. Loss of balance may be the only symptom with a tendency to move to one side as in case D.S. (page 46). Case No. 348 felt he was tipping over backwards.

Case No. 348 for his two years of diving always had pain in his ears on descent and had pain in his ears on ascent. At times he developed vertigo on ascent. His vertigo was unique. He felt he was tipping over backwards. He was treated for the nasal congestion of his vasomotor rhinitis, by cautery of his inferior turbinates. He was counselled on the technique of clearing his ears and given jaw muscle stretching exercises. In addition his reflux indigestion and lump in the throat syndrome was also treated. The cautery cured his vertigo.

A further variety of vertigo is when the movement of the eyes is slowly from side to side or slowly up and down - the flutter or shimmer effect. This is an early form of vertigo and is called oscillopsia.

Case No. 327 suffered from dizziness on ascent but his description was that objects moved from side to side, back and forth and not in a spinning manner. This is oscillopsia.

Case No. 366 had vertigo on ascent with a pressure in his left ear. He said his head was see-sawing and spinning. His vertigo was less with his eyes closed.

The cause of this vertigo is easy to understand. It is linked to the balancing apparatus or the labyrinth of the inner ear and the connections to the **Balancing Centre** in the base of the brain. From this Balancing Centre there are nerves joining it to the muscles which control eye movements. These connections are a necessary part of posture and balance when turning the head. The balancing apparatus can be likened to a two-engined aeroplane in that each engine has a tendency to pull the aeroplane towards the opposite side. This is precisely what is happening in the ear.

The nerve impulses cross over and the left balancing apparatus tends to pull the eyes to the right and the right side pulls the eyes to the left. Normally the pull is equal and the eyes remain stationary (Fig. 27).

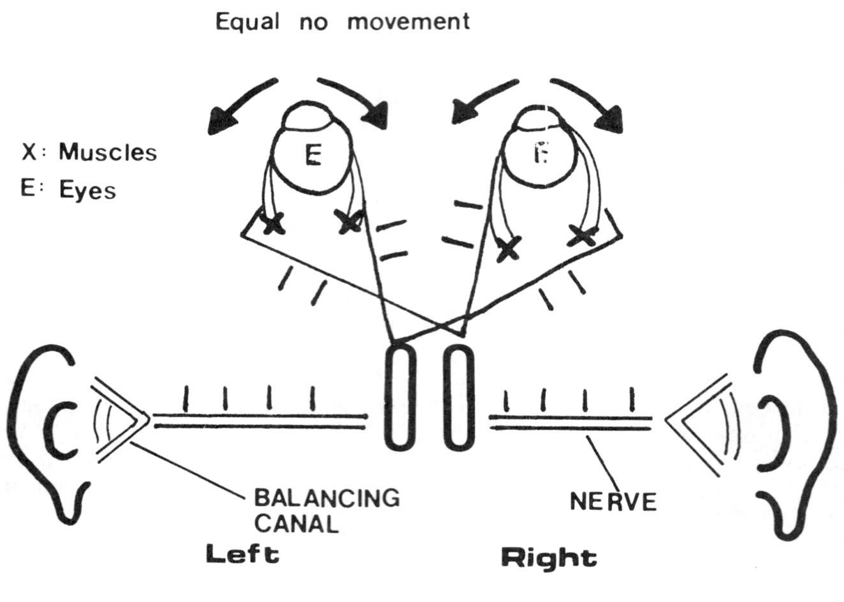

Fig. 27

The impulses are equal from each ear, so there is no movement of the eyes caused.

When a diver is ascending, according to Boyle's Law, the air inside the middle ear cavity expands, in fact it doubles in size for every 33 feet (10m) of ascent. Normally the extra gas volume escapes down the Eustachian tube as the pressure in the ear rises relative to the surrounding water pressure. If this air does not escape then it pushes the eardrum outwards to such an extent that in some cases pain is felt. If there is a weakness in the eardrum, as a result of ear trouble as a child, the eardrum may burst or rupture and the air can escape. However, at the same time, the increasing air pressure pushes the stirrup bone (stapes) into the inner ear (Fig. 15, p. 45). It is that part of the inner ear that forms the balancing apparatus that is right next to the stirrup bone. The irritation by the stirrup bone causes a few more nerve impulses to pass up from the ear affected (Fig. 28). It occurs in one ear, because as most divers are aware, they have one Eustachian tube which clears easier. The two Eustachian tubes vary in their ability to open.

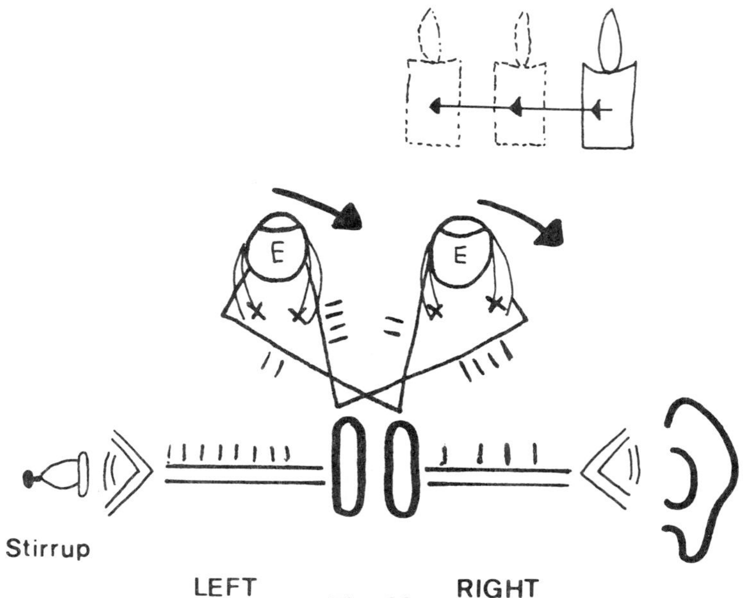

Fig. 28

Stimulation of the left balancing apparatus causes more impulses, from the left ear, making the eyes move to the right and the candle seems to move to the left.

So if these impulses arise from the left ear then the eyes unknowingly turn slowly or revolve to the right (Fig. 28). The conscious brain corrects this movement, so the eyes are quickly moved back to the central position. The conscious brain relaxes, the eyes again move to the right and this is again corrected. So the net effect is that the diver appears to see objects in front of him move across to the left (Fig. 29). As the number of nerve impulses from the left ear increase so the speed of movement of the eyes increases until if you look at the eyes you will see them moving from side to side or flickering. This movement occurs slowly one way and quickly the return way. This is the reason for the spinning sensation and the disorientation. The balancing apparatus moves the eyes slowly, in this case to the right and whatever is seen by the diver, who does not know his eyes are moving, is moving to the left. The very quick recovery (Fig. 29) of the eye back to the central position is not noticed and the next thing is that as the eyes move slowly to the right again. The objects again seem to be moving to the left. As the repeated movements just show objects continually moving to the left the diver has reason to believe that his surroundings are moving continuously to the left in a circular manner. Hence, the end result is an hallucination in which the surroundings appear to be spinning from right to left, in other words **vertigo.**

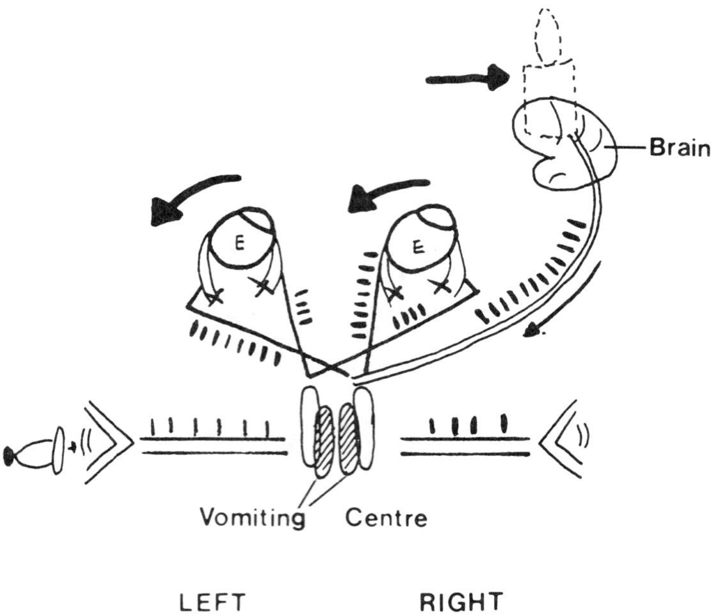

Fig. 29

The thinking part of the brain steps in to move the eyes back to the front and this movement occurs very quickly.

Because the balancing centre of the brain is situated next to the vomiting centre and because sudden activity of one group of nerve cells causes a radiation effect to those next door, the vomiting centre is stimulated. This may be just a feeling of nausea but it can advance to vomiting. This effect of the radiation from the balance centre to the vomiting centre is the cause of motion sickness or sea sickness. In this case, because of the big increase in impulses coming from the balancing apparatus, as a result of the unaccustomed motion of the inner ear the susceptible subject becomes nauseated, unwell and may eventually vomit.

As the input is equal from each ear no vertigo occurs. However, it is possible by conscious effort and concentrating, for the person to be able to overcome all these effects. If the person does not just resign himself to the vertigo or vomiting but tries to fix his eyes on one object then the vertigo or the sea sickness can be overcome, but maybe only temporarily. Sometimes tilting the head back causes a different balancing canal to be affected and the nausea passes off.

It is not only the pressing in of the stirrup bone in the Alternobaric Vertigo that causes vertigo in divers. If the left eardrum ruptures and cold water gets into the middle ear, the water cools the left balancing apparatus and thus reduces the number of impulses and so the right side over-reacts and vertigo occurs in the opposite direction, with objects going from left to right.

Case No. 627. This was a lady of 25 years who was down to 10 feet (3m) depth when she felt pain in her left ear. There was a loud pop and the pain suddenly stopped. Vertigo with objects going left to right occurred so she gave up diving and reported for examination two days later. She had a perforated eardrum and a deviation of her nasal septum to the left. Antiseptic (iodine and boracic acid) powder was blown in and she did not attend again. It was a thin drum which ruptured at 10 feet (3m) and let water in causing a slow movement of the eyes to the left and a quick component to the right.

If the Round Window membrane ruptures due to prolonged nose-holding and blowing hard down the nose then the balancing apparatus can be mildly upset. One patient with a ruptured Round Window membrane only complained of the fact that on arrival at the wharf that she couldn't walk straight up the gangway. She had a tendency to veer to the right or because her eyes moved to the left, she saw the surroundings moving to the right so she tried to keep up with them.

Case No. 364 was a 37 year old male diesel mechanic with 24 years of diving experience. He had a cold in the nose and an alcoholic hangover when he went for a dive in a rough sea. He did a quick descent having difficulty in clearing his ears and having pain in both. He had pain under his eyes on descent. He developed vertigo on ascent and when he surfaced he felt blocked in the right ear. About four hours after the dive he noticed a ringing noise or tinnitus in his right ear. He did not notice the right deafness until the day after the dive. Two days later he was examined and put into hospital for treatment of his severe 75 db inner ear deafness. He showed no improvement and underwent operation for repair of the Round Window rupture nine days after the dive. This gave him better hearing on the right (average 30 db) than on the left with both ears probably slightly deaf from his noisy occupation. Two months after the operation he went back diving once every one or two weeks promising to come and see me if he had any trouble but 20 years later he has not come back.

Case No. 357 was a 38 year old male who had been diving for 19 years and was a professional diver. He was seen one week after treatment in the decompression chamber for decompression sickness. He had been doing a helmet-dive to 120 feet (35m) and was at the bottom for 86 minutes. Although the usual decompression stops were made he eventually ended up in the chamber. Of note was that his symptoms did not begin until about 1 1/2 hours after the dive when he developed vertigo, nausea and vomiting. This stopped once he was under compression.

In white-out conditions or when standing on an underwater cliff-top, because there is nothing to fix the eyes on, this can result in the eyes oscillating from side to side and this produces vertigo.

Summary

The eye movements are influenced by the balancing apparatus of the ear through the balancing centre in the brain. The left ear moves they eyes to the right and the right ear moves the eyes to the left and normally there is an equilibrium in this system. An increase (Alternobaric Vertigo) or a decrease (Round Window rupture, cold water into the middle ear) in the impulses causes an unrecognised movement of the eye to one side and a very quick return, thus the person is aware of objects moving but in only one direction and hence they have the hallucination of vertigo (See also Chapter 5).

CHAPTER 11

THE JAW AND UNDERWATER SWIMMING

In treating scuba divers for various complaints it is manifestly almost impossible to go scuba diving without suffering earache. In fact, pain in the ear is used by the majority of divers as an indication for them to inflate the middle ear to equalise the pressure. However in many scuba diving patients the complaint of pain appears to be excessive and not in keeping with the mild ear pain used as a warning signal by the average diver. In addition some complaints of earache were not associated with the expected pressure change and some ear pains persisted after diving had ceased. The normal way in which the teeth set together (occlusion) is upset in divers, because they are clenching a mouthpiece which interposes at least ½ inch (6mm) of rubber between the teeth thus bringing about a form of malocclusion. An investigation was carried out to see how this problem affected underwater swimmers.

In examining 322 consecutive scuba divers reporting with various complaints, questioning was carried out to determine whether any of the signs, symptoms and effects of the Jaw Joint Disorder were present (Table 3). Questioning was also directed to whether these divers had any of these effects when they were not diving.

Table 3
Jaw Complaints of 87 Divers

Excess ear ache	41	Low tone deafness complaint	13
Blocked ear feeling	28	Temple headache / pressure	11
Noises in the ear	12	Neck / throat soreness	5
Dizziness	9	Dental disorders discovered	110
Popping, clicking	5		

Thirty-nine divers had this disorder as their main condition or diagnosis (Table 4), 32 divers had this disorder as a secondary condition and so on, so that of a total of 322 diving patients 87 or 27% of them suffered with the Jaw Joint Disorder.

Table 4
87 Scuba Divers with the Jaw Joint Disorder

Main diagnosis	39
Secondary condition	32
Third and fourth condition	16

Table 5 shows the details of the 110 dental disorders, the main being teeth worn smooth from unrecognised grinding, a history of teeth grinding and wisdom tooth trouble.

Case Histories

G.S. who was diving three weeks previously to a depth of 30 feet (9m) when he had no trouble clearing his ears even though he had a "bit of a sniffle". He was suffering from an uncomfortable left ear which throbbed at times and sometimes felt numb. He had a sore throat on the left on swallowing and the soreness radiated to his left ear. Although he was unaware of grinding his teeth, the front teeth were worn flat. He did admit to being a tooth clencher and to biting his mouth piece hard. The shape of his eardrums indicated spasm of his middle ear muscles. The sequence of events was that he had minor Jaw Joint Disorder pre-dive, but the excessive muscle tension incurred whilst biting his mouthpiece plus the cold of the water set off the odd feelings in his ear. Once these feelings developed they persisted as the muscles did not lose their tension. Jaw muscle stretching exercises cured this problem.

Table 5
110 Dental Disorders Discovered

Worn teeth	25	Jaw joint clicks / aches	17
History of teeth grinding	15	Missing molar teeth	8
Impacted wisdom teeth	12	Ill-fitting teeth	8
Erupting wisdom teeth	5	Teeth ache	2
Others	8		

Case R.H. would complain of a hearing loss with blocked ears and dull feeling in his head despite ease of clearing the ears. Suddenly he would hear a click in his ears and all these symptoms would pass off. He had a history of clenching and grinding his teeth and had indrawn eardrums. It was taken that the audible relaxation of his middle-ear muscles resulted in the clearance of his blocked ears.

Treatment

In the majority of cases in which there is no great dental abnormality an explanation of the cause and the exercises are usually sufficient treatment. Reassurance that the disorder is not related directly to his diving enables the diver to carry on without any concern for the symptoms which arise. Most of the divers come because they feel that the symptoms could possibly lead to some complaint directly related to their ears and hearing. For those cases with obvious dental abnormalities, referral to the appropriate dentist or dental specialist is indicated. Impacted wisdom teeth obviously have to be removed, ill-fitting dentures should be replaced and the dentist can help by grinding down any tooth cusp that projects too much. This is called “removing interferences” so as to “balance the occlusion.” For those who find themselves biting their mouthpiece firmly, the advice to hold it loosely with their lips, is also helpful.

For those who grind their teeth excessively, if an explanation and exercises are not sufficient, a short five night course of chlordiazepoxide 5mg at bedtime seems to produce a long-lasting effect. It is suggested that once the habit has been broken by the use of this tranquilizer and the muscle stretching exercises are carried out, the overall subconscious need to grind and clench the teeth is reduced to normal limits. By opening the mouth as wide as possible the jaw muscles can be stretched and this reduces muscle tightness reducing the pain signals. Done regularly before meals and on wakening in the morning jaw muscle stretching exercises by themselves are often sufficient to relieve the problem.

JAW JOINT DISORDER EXERCISES

This sheet is for your information. It is to show you how to overcome some, if not all, of your troubles.

The average person does not know much about the jaw joint. There is no folk-lore or folk-medicine about it. It is recognised that patients have some difficulty in understanding their problem. Some patients do not believe what they are told.

Examination and experience has shown that there is no disease present. Any aches, pains or other feelings can arise from a disorder of the normal jaw muscles. They may only be felt in the ear.

Grinding your teeth and clenching your teeth together are harmful habits and must be stopped. Your teeth should never be together. This can cause muscle spasm which then goes onto pain. The pain may be felt in the muscles under the eye or in the front or side of the head or just in the ear.

It is important to use the teeth on both sides of your mouth when chewing food. Gentle chewing from side to side is natural and necessary.

When chewing food, keep your lower jaw as far back as possible.

If you have dentures, ensure you wear them 24 hours a day. They are needed to keep your jaws apart.

EXERCISES

1. Gentle opening and closing movements of the mouth can be practiced on waking, before meals and at bed-time.

Increase the movements gradually, to eventually open as widely as possible, but not to the extent of causing any pain.

2. To do this, hold your chin between your forefinger and thumb, apply mild pressure backwards, and slowly open your mouth, keeping the lower jaw pressed back as far as possible. Take at least three seconds to open your mouth.

3. After holding your mouth open for 3-5 seconds, slowly close it. Pause, then slowly open it wide again. Repeat this opening and closing exercise six times at a time, 2 to 4 times daily.

4. You should also gently stretch the muscles in the neck with three different exercises. Move the head from side to side, forward and backwards then turn it to the left then to the right. Twenty moves each way are indicated especially when you are warm, for instance after a bath or shower. Neck tension affects jaw muscle tension.

NOTE

1. The common mistake is to open mouth too quickly.
2. Never carry out the movements sufficiently great as to produce pain or discomfort.
3. Do the exercise in front of a mirror at first to ensure that you open your jaw straight up and down and with no jerking movements.
4. Symmetrical and even movements of the jaw are needed.

Case History

With thoughts of the disorder in mind after a recent dive, diver **N.R.** found himself unable to bring his incisor teeth together to bite his lunchtime bread roll. He turned to his companion, an expert in underwater medicine and asked him whether he had any trouble bringing his teeth together. He said "Yes" and went on to say that after diving he found that he had an ache from his ears radiating down under his jaw. In neither case were these symptoms sufficiently severe to cause any trouble and within 30 minutes the muscles had re-adjusted themselves from the forced-open position due to holding the mouthpiece with the teeth, and a normal bite position was achieved. However if there is any pre-dive inclination towards this disorder then the symptoms can be worse.

Pain

There are various reasons why pain is referred to the ears from the jaw joint and the jaw muscles. The back wall of the jaw joint is the front wall of the outer ear canal (Fig.30). In those situations where the pain is sub-threshold (there are not enough pain endings stimulated to actually feel pain), then it may be felt as an itch in the ears. Ear pain may then develop with minor pressure changes due to the additive effect of stimulating nerve endings plus the influence of cold on the muscles. In this latter state, pain can be in front or behind the ear and radiates down the neck or along the jaws, a distribution rarely described in those who actually rupture their eardrums. That pain is deep in the ear, comes on gradually, builds to a climax and suddenly reduces when the eardrum ruptures.

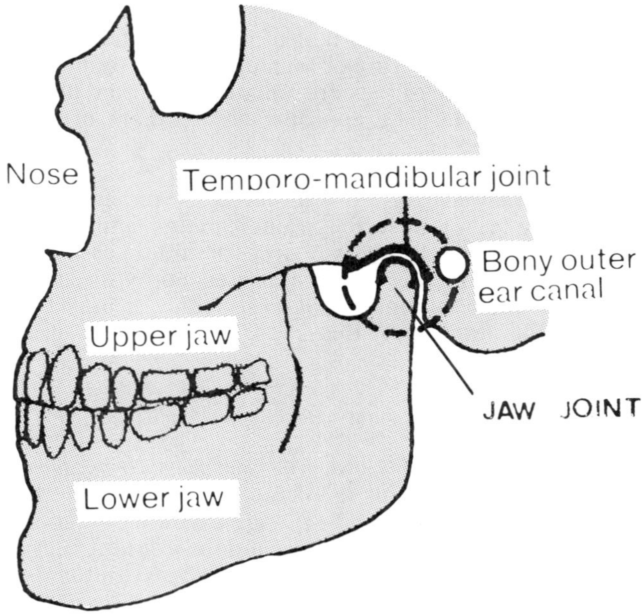

Fig. 30
Jaw joint or temporomandibular joint is immediately in front of the outer ear canal.

Blocked Ears

There is evidence to show that the low tone deafness and blocked feelings in the Jaw Joint Disorder are due to the nerve connections between the jaw muscles and the middle ear muscles. When the middle ear muscles contract, as they do in those who grind their teeth, it stops the eardrum from vibrating freely by pulling it inwards. This causes the minor low tone deafness and the blocked feeling.

Noises in the Ear

The popping and clicking sound that five of the divers complained of could be due to sudden contractions and relaxations of the middle ear muscle or to misplacement of the cartilage in the jaw joint. It is not always noted by the patient to be arising from the joint but is stated by them to be in the ear.

That ringing noises can be caused by the Jaw Joint Disorder is demonstrated by the fact that treatment of the disorder results in a disappearance of the noises. Such noises are not the usual high-pitched hissing sound or sound similar to that of grasshoppers but is more often described by the patients as a ring or hum or buzzing sound. In such cases there is noted an indrawing of the eardrum due to marked tension or spasm of the middle ear muscle. The appropriate treatment often results in a disappearance of this noise.

In dealing with scuba divers it is important to remember the Jaw Joint Disorder when the diver describes aches, pains and other symptoms referable to the ears. It is often mixed up with middle ear barotrauma (the squeeze), the most common disorder experienced by divers. Unless this other common disorder is considered every time, it may be difficult to explain many of the complaints that cause concern to divers, and wrong advice may be given.

Case History

Case MDS. This 34 year old male with a two year experience of diving attended because of pains in his temples on diving. He would stay at one level and the pain would pass off and he would descend further until he developed pains again in the temples and he would rest again. He had a history of some blockage of his nose but he had no trouble in clearing his ears. He had a tendency towards biting his mouthpiece hard, but no history of grinding or clenching his teeth. He had clicking in his left jaw joint at times. His eardrums were retracted and his nose was a little narrow. He had marked wear of his front teeth from grinding them.

The diagnosis was made of the Jaw Joint Disorder whereby tension in his jaw muscles aggravated by the cold water produced pain in his temporalis muscle which covers the temple. However he reported back in three months time stating that he was still suffering from pains in the head on descent and also pains in the head on ascent. He apparently had not been carrying out the jaw muscle stretching exercises that he had been advised to do and so this was re-emphasized.

This same patient reported back six years later stating that all his problems had been settled with the jaw muscle stretching exercises. However this time he was complaining of a buzzing noise in the left ear which he variously said began in December, then January, then February and then 24th January. He stated that he had had left deafness in mid-

December after diving and this deafness was worse the next day. He had been diving to 35-40 feet off a boat. He still had had no difficulty in clearing the ears. The day after his dive in mid-December his balance was upset and from that time onwards he had, when walking through a doorway, hit the edge of the door three or four times but equally left or right. The deafness noted in mid-December lasted for a month and gradually improved. He had a buzzing noise in his ears which came and went and he only heard it at night. He stated that he was now clenching his teeth less and was no longer biting his mouthpiece but he still had a clicking left jaw joint. An audiogram showed that he had a minor high-tone hearing loss in both ears which could not be related to his complaints.

Again the diagnosis of the Jaw Joint Disorder was made and he was put onto jaw muscle stretching exercises. He went away quite happy but returned three years later with a history of a whistling noise in his ears for four months.

This was not related to diving at this stage. He had now developed the "globus hystericus syndrome" in which he was complaining of a lump in the throat and this was typical of his general anxiety state which of course is part of the Jaw Joint Disorder.

A reasonable explanation can be given in this case. To understand this type of diver a knowledge of this disorder is important. Pains in the temple are common but to relate them to descent and ascent is difficult. With descent the water temperature usually declines and this could have been a factor. As he was clearing his ears he could have been biting his mouthpiece hard, stressing his temporalis muscle at intervals. On ascent he could be thinking about his pain as at this phase in diving most divers relax and float slowly upward with not much to occupy their minds. He talked more about pains in the head at his second visit and these could have been a combination of temple and ear pain. The latter would be due to expansion of gases on ascent resulting in stretching the eardrum made sensitive to pain or put into a sub-clinical pain state by the Jaw Joint Disorder.

His complaint of buzzing, deafness, and balance upset could be pressure-related but in this case they were put down to this disorder. Blocked ears are often described as deafness. When a normal audiogram is obtained one can persuade the diver that their deafness feeling is really due to a blocked feeling. The blocked feeling being due to spasm of the middle ear muscle causing eardrum indrawing.

The psychological aspect of this disorder is what makes the symptomatology so varied in intensity. Thus a complaint of deafness can be decreased in severity to a blocked feeling. Such people tend to exaggerate their symptoms for fear that they will be overlooked. In fact they do often have barotrauma, but part of their problem is based on the Jaw Joint Disorder. Thus two forms of treatment are often required.

CHAPTER 12

THE SCUBA DIVING DOCTOR

Introduction

The first eleven chapters of this book were written for scuba divers in order that they may understand more about how they function, both physiologically and when under pressure, and how they can care for their ears and nose. Throughout every effort has been made to use terminology that they can understand but some medical terms they must learn, such as the Eustachian tube. Accordingly, in places oversimplification occurs and descriptions may lack depth of detail. It was considered advisable to add a chapter directed towards the scuba diving doctor. In this chapter terminology will be according to medical standards and may not be too clear to non-medical readers, but then this may be underestimating the intelligence and knowledge of experienced scuba divers who still wish to understand more about themselves. They are invited to read on.

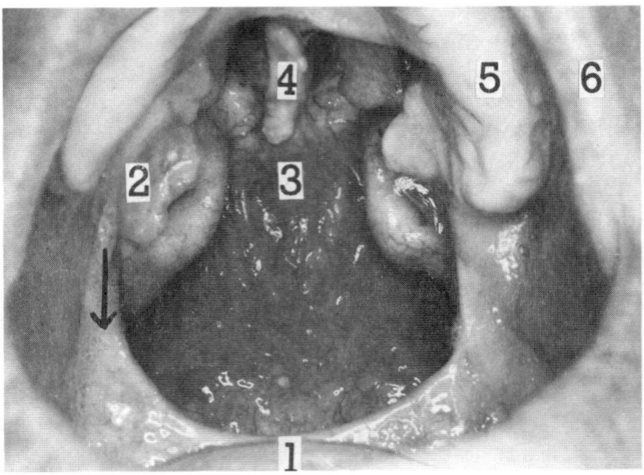

Fig. 31

Photograph of the Eustachian cushion (2) through the mouth with the soft palate muscle passing downwards. In this patient all of the bony hard palate has been removed. The tongue (1) is against the soft palate, (3) is the region of the nasopharynx, (4) the nasal septum, (5) the gum of the upper jaw and (6) the left upper lip.

THE EUSTACHIAN TUBE

With the more intensive investigation of this region by those delving into the causation and treatment of secretory otitis media ("glue ear") more information is coming to light. It is now acknowledged that there is no direct relationship between the size of the adenoids and the amount of Eustachian tube blockage. In this context "Eustachian tube blockage" means that the normal opening of the tube is more difficult to obtain. There is a relative loss of normal function or it can be called Eustachian tube dysfunction. When the diver is living on earth's surface, it works quite well but when increasing pressure occurs on diving, difficulty develops in getting air to pass up and sometimes down the Eustachian tube. Normally the tube is closed or blocked and it only opens with swallowing. The function of the Eustachian tube may be normal even with large adenoids present and removal of the adenoids does not necessarily relieve Eustachian tube dysfunction with its secondary effects on the middle ear. Some scuba divers have adenoids and have no difficulty in clearing their ears.

When the rare case of **Patulous Eustachian** tube is seen, because their Eustachian tube is open all the time, they make ideal divers. However paradoxically they usually complain of blocked ears. They also can often hear their own voices inside their head or can hear their breath going in and out. The eardrum can be seen to move with a sniff or a nose blow.

If allergic rhinitis is present then often there is no change in the Eustachian tube blockage after adenoidectomy. The nasal congestion of any form of rhinitis, be it allergic, catarrhal or the Common Cold, is the greatest cause of Eustachian tube dysfunction in an adult. This is the reason for advocating and for the success of nasal treatment in relieving those divers with difficulty in inflating their ears. In testing prospective divers for Eustachian tube function, a positive result of the test described by Joseph Toynbee in 1873 is presumptive of normal function. In this test the diver holds his nose and swallows. If he hears a click or a pop or feels the pressure rise in his ears, then his tubal function is satisfactory for scuba diving. If he fails after two consecutive swallows then his tubal function is probably defective. In those who fail Toynbee Manoeuvre or who have difficulty in inflating the middle ear the first approach is to examine their technique (Chapters 2, 3 and 9), then follows treatment of any of the many forms of rhinitis or nasal congestion.

Cautery or heat treatment of the inferior turbinates is often sufficient. This should be done before advising the diver to have a septoplasty. If he does have a septoplasty it is important to have a radical procedure to include removal of the posterior vomero-ethmoidal spur.

Teaching Technique of Clearing the Ears

The usual procedure is to tell the person to perform Valsalva Manoeuvre but this does not always work, because with no pull on the levator palati muscle, the raised air pressure merely compresses the inner end of the Eustachian tube, locking it. The person has to originate some form of stretch by the levator palati muscle. This can be done by teaching how to feel when this muscle is stretched as when the palato-glossus muscle pulls the palate downwards.

At the beginning of a yawn the Palato-glossus muscle contracts and the diver should be told to practice the beginning of a yawn and with experience he will be able to feel this muscle in action. By trying to imitate this feeling the ability to contract this muscle can be developed. The feeling occurs inside the angle of the jaw, at the back of the mouth.

With the expert blower of smoke rings, the main propulsion of the smoke is brought about by contraction of the Palato-glossus muscle and the muscles of the floor of the mouth such as the Mylohyoid. Again once when the feel of this is recognised it can be imitated.

It is possible to hold a mirror in front of the mouth with the back to a source of light, such as a window. The light is reflected into the mouth and the palate is readily seen. When this muscle contracts the soft palate descends slightly but more obviously the anterior pillars of the fauces, the fold encasing the Palato-glossus muscle, moves towards the middle. The posterior pillars of the fauces (Palato-pharyngeus muscle) also moves towards the mid-line. An experienced diver can demonstrate this and his pupil can then use the visual coordination to mimic this movement.

With some divers the fact that their mouths are propped open with the mouthpiece makes this movement more difficult, but practice out of the water with the mouthpiece can be carried out until clearing the ears becomes easy.

THE MIDDLE EAR

If, in the examination of the eardrum, a Siegle's pneumatic speculum is used, usually part of the auroscope diagnostic set, fewer mistakes in diagnosis will be made. On placing a Siegle's speculum (Fig. 32) in the ear, not only is the drum magnified in size by the lens, but also, by intermittent pressure on the rubber bulb, the eardrum can be made to move to and fro, imitating the vibrations necessary to transmit sound waves. The production of visible movement of the eardrums gives much more information to the examiner. Any disease in or around the drum can prevent it from moving freely, resulting in reduced hearing. For example, the collection of wax on its outer surface, the accumulation of fluid in the middle ear, or a perforation can prevent

Fig. 32

Siegle's type pneumatic speculum with magnifying lens and rubber bulb for producing eardrum movement. The presence or absence of movement supplies important information.

this movement. If the drum is too thin, it will move excessively. When a normal eardrum ruptures due to excess pressure in the outer ear canal, it usually does so just posterior to the handle of the malleus, in a linear fashion. In an eardrum where there has been loss of the fibrous middle layer, it will occur in this thin area. Such a perforation becomes more obvious when the drum moves and a potential slit-like perforation may have its edges forced apart when using the Siegle's speculum.

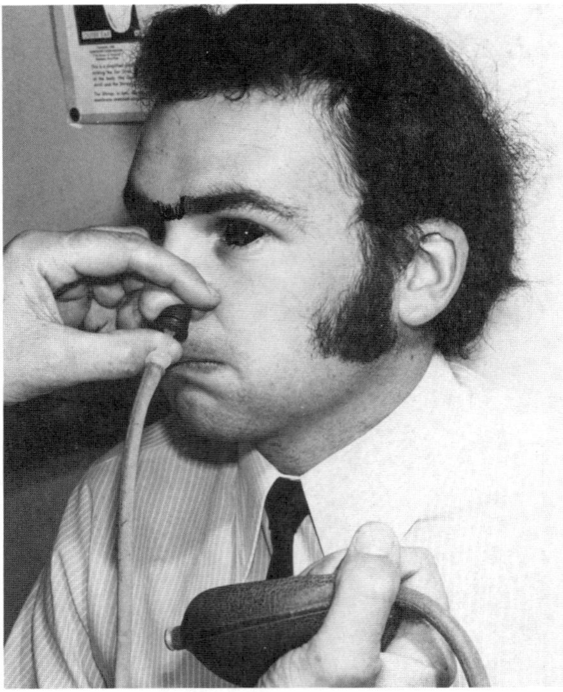

Fig. 33

Politzerisation. With the rubber nozzle in one nostril and the other clamped shut, the rubber bulb is suddenly compressed and air is forced up the Eustachian tubes.

When the tear in the drum is in a linear fashion, after pressure equalisation the edges come together so that perforation is not visible. In 48 hours it can be healed but the reddened scar will still be visible for up to 10 days. With a ruptured eardrum no deep divingis allowed until this heals as it usually does spontaneously. While waiting for it to heal the outer ear canal should be kept dry and sterile with antibiotic powder. Most important is to keep water out, and if he is to dive, shallow diving with a paraffin ear plug can be allowed (see Chapter 4). If the perforation is large or persists for more than three months or recurs with minimal pressure changes, then a myringoplasty is indicated. After a repair of the eardrum perforation, the eardrum is usually stronger and thicker than before the operation. It takes at least two months from the time of such an operation for the ear to heal sufficient for diving.

POLITZERISATION

This is a method of forcing air up the Eustachian tube. The rubber nozzle of the Politzer apparatus is inserted into one nostril, the other being blocked, and whilst the patient is puffing out his cheeks with the lips tight together, the rubber bulb is suddenly compressed. The patient will hear a hiss, pop or bang, or will feel pressure in the ear. Alternatively the rubber bulb is suddenly compressed at the moment the larynx rises during a swallow when the palato-pharyngeus muscle is pulling on the palate which pulls on the Eustachian tube.

OUTER EAR

The collection of wax in the outer ear canal is a self-inflicted condition (see Chapter 4). When a person has an itch in the ear or if it feels blocked, relief is usually obtained by poking it or rubbing inside, in front of, or behind the ear. The prime cause of these ear sensations can be due to a disorder of the temporo-mandibular joint with the reaction of the jaw muscles going into tension in trying to correct this. The tension in the jaw muscles may be due to overuse in those who grind or clench their teeth excessively. As mentioned previously, earache can be caused this way, but before pain occurs, an itch may be present. This itchy ear can be relieved by rubbing Eurax lotion (Geigy) just inside the opening of the outer ear canal. The person can then be asked to leave his ear alone to allow the normal exodus of the wax. Relief of the blocked sensation, due to the tensor tympani muscle going into sympathetic spasm with the overused jaw muscles, is obtained by dental treatment such as removal of interferences. Rubbing the outer ear canal opening also causes the tensor tymponi muscle to relax. People think that this rubbing moves the piece of wax blocking the ear so they keep rubbing. However this pushes the wax back in. The person is told to give up grinding and clenching the teeth and biting the mouth-piece hard. Jaw muscle stretching exercises are also helpful (Chapter 11). Failing this, after explanation, the diver can be told not to interfere with the ears, even if they feel blocked. The diver must learn to differentiate this blocked feeling from the blocked sensation due to the pressure of the water on descent.

Wax Removal

The wax may be removed by small wax hooks or curettes or with aural alligator forceps. The wax may be softened by instilling Cerumol drops or olive oil. Note that Cerumol has a turpentine base and if given to anybody with external otitis, causes intense irritation and pain. Syringing the ears is the commonest way of removing wax and is still an efficient procedure for those people who do not have perforated eardrums. Repeated gentle syringing is better than one forceful syringing. Syringing before the ear has been inspected should be avoided. One should not syringe an ear from which there is any discharge, nor an ear that is aching or painful.

Dangers of Syringing the Ears

(1) Leaving water in the ear canal.
(2) The presence of a perforated eardrum.
(3) Loose nozzle flying off the syringe.
(4) Damage to ear canal.
(5) Forceful syringing.
(6) Syringing before the ear has been inspected.
(7) Using cold water which then causes vertigo.

Treatment of External Otitis

The cleansing of the skin is of paramount importance in order to allow local applications to come into contact with the inflammed skin of the canal. Self-made thin cotton-wool tipped probes are the most practical although the use of suction is the easiest. Fine suction nozzles can be used through an auroscope so that the cleaning is done under direct view. The commercially prepared cotton-wool buds are too big, too hard and too rough. The skin of the ear canal is very thin and easily damaged. The commercial cotton buds because of their size and consistency tend to push wax and debris down on to the eardrum. As they are relatively hard and rough they damage the delicate thin skin of the outer ear canal, causing more problems.

A wisp of cotton-wool wound on to the end of an orange stick, a Jobson-Horne probe, or even a match stick is best. At least 5 mm of cotton-wool should project beyond the end of the stick. It can be used blindly to clear the outer half of the ear canal then through an auroscope to clean the inner half. The patient is told to tease out a piece of cotton wool about one inch (2.5 cm) long and place the tip of the match stick two thirds the way along the wool held with the left thumb and forefinger. The stick is then turned and the cotton wool pressed against it. The cotton tipped stick is then twisted the same way into the ear canal. To straighten the ear canal and for ease of insertion of the probe, the pinna is pulled up and back.

The most effective and safest way to clean out the ear canal is the use of suction through an auroscope so that the tip of the metal suction tube is continuously in view.

For most cases of outer ear infection or external otitis the following three types of eardrops after adequate cleaning will cure the problem. The instructions are to use Bottle A for four days, then Bottle B for four days then Bottle C for four days. Do not use all the drops at the same time. If any drops sting go back to Bottle A, the healing drops, and start the four day cycle again. When you can use the "C" drops (the Spirit drops) for four days without stinging you are cured. Each lot of drops are stronger, help healing and toughen up the skin to prevent relapse. However the most important feature of the treatment is to clean out the ear and never use commercial cotton buds.

Bottle A: Soframycin, gramicidin and dexamethasone eardrops
Labelled: Bottle A. 3-4 drops in affected ears three times daily after mopping.

Bottle B: Glacial Acetic Acid 2% in Vosol
Labelled: Bottle B. 2-3 drops in affected ears daily.

Bottle C: Glacial Acetic Acid 5%
Distilled water 20%
Isopropyl Alcohol to 100%
Labelled: Bottle C. 2-3 drops in affected ears daily.

Note that the distilled water can be increased in Bottle C if it burns or stings the normal skin. Bottle C, the spirit drops are very good for prevention. See Case History on page 30. However normal untouched ears should cope under all circumstances with diving and swimming.

INNER EAR

Hair Cells

The Hair Cells also known as Hearing Cells are the key for understanding inner ear deafness and dizziness. These cells have hair-like processes projecting into the tectorial membrane of the cochlea. When these hairs are bent by the vibrations of the membrane or by flow of the fluid, a nerve impulse originates from the body of the cell. They are easily damaged but can often be revived by treatment. The main treatment is to increase the local oxygen supply and to use the anti-stress corticosteroids such as prednisone or dexamethasone.

Rupture of Round Window Membrane

When a person performs Valsalva Manoeuvre there is a transient rise in blood pressure but this quickly drops to normal. However, with the raised intra-thoracic pressure there is decreased venous return resulting in the filling of the veins in the neck and then the head, so that their venous pressure rises.

This rise in pressure prevents the re-absorption of the cerebro-spinal fluid (CSF) from the cerebro-spinal fluid circulation so that this pressure rises in turn. There is a prolongation of the subarachnoid space into the base of the petrous temporal bone forming the **Aqueduct of the Cochlea.** This opens into the perilymphatic system adjacent to the Round Window. In people in whom this canal is larger than usual, this raised CSF pressure is transmitted immediately to the perilymphatic canal whose lower end is blocked off by the Round Window membrane. This thin membrane is then subjected to pressure and it can rupture with an outflow of CSF and perilymph into the middle ear. The symptoms of this are variable but any sudden onset of hearing loss, with or without balance impairment or vertigo can be caused by rupture of this membrane. With small ruptures the hole can seal itself with the restoration of normality, but in many cases the loss of perilymph leads to loss of function of the Hair Cells and eventually cell death. Operation within one week to repair this easily accessible membrane is usually followed by complete restoration to normal of the hearing and balance. Such patients should then be warned of the likelihood of a recurrence. They should be counselled on the simple methods clearing their ears. They should be capable of clearing their ears easily, or the person should give up scuba diving. So far there have been no recurrences of this rupture seen in affected divers that have been treated or examined. But this is not universally the result as in some cases the repair fails to heal or they suffer a second rupture.

Treatment of Sudden Deafness

Bearing in mind that the pathology of most types of sudden deafness is damage to the Hair Cells, lack of oxygen supply to the Hair Cells or to an decreased amount of supporting perilymph fluid, then the following treatment is obvious. Prednisone mg 20 three times a day, hydrochlorothiazide mg 25 twice a day, and Carbogen therapy, either at atmospheric pressure or in a compression chamber at two atmospheres for a period of one hour twice a day. The best place for treatment of any sudden deafness in the inner ear is in hospital. Sometimes it is only the high frequencies that are affected and this can easily be determined in most scuba divers by testing their hearing with a wrist-watch. The high-pitched tick will not be heard in an ear with the usual high-tone loss, which is often described by the diver as a blocked sensation, mostly accompanied by high-pitched noises in the ear. To keep the CSF pressure low, hydrochlorothiazide, bed rest and no physical exertion is required. Intravenous or oral oxypentifylline (Trental, Hoechst) is the latest addition to treatment as this increases capillary blood flow and thus the local oxygen supply.

All cases of sudden sensorineural deafness should be treated medically for at least 3 days before surgery is an option. If the patient's symptoms become worse then an exploration of the Round Window membrane is advised. If the symptoms remain the same it is after 6 days of medical treatment that the surgical option is considered. Many cases heal and recover spontaneously with medical treatment.

Case History 948. This diver had difficulty in clearing his left ear on descent to 40 feet (13m) in a slow descent and a slow ascent. It was his first dive in 3 years. On surfacing he immediately noted left deafness and a low pitched tinnitus which then became a ringing sound. He had a loss of balance on the boat and at least once each of the next 3 days he nearly fell over. I saw him on the 3rd day and immediately admitted him to hospital for routine treatment of sudden barotraumatic sensorineural deafness. The audiogram was 65-75db down for the whole hearing range. After 4 days he was still deaf and had some vertigo so he was operated on the next day. An obvious ruptured Round Window membrane was repaired and his balance returned and his hearing became normal.

Collected Doctors' Remarks

On a lighter side of divers' problems is what they report their family doctor to have said. Undoubtedly these are not always correct but are the diver's recollection influenced by his understanding. When a diver told his doctor that he had a mouthful of blood after a dive he was told' "You have burst your top sinuses. Don't worry. It will heal in a couple of hours." Another quote was, "Your ear looks like a vacuum-packed chicken and if you continue (to dive) you'll rupture it." Another diver with a cold was told, "You've probably got a bit of muck up there and it will take 3 or 4 weeks to go." (What about a bit of decongestant treatment?). Unfortunately such statements besides being untrue tend to frighten the diver and they then come to see me. The first was an ordinary nose bleed 98% of which come from thin-walled veins in the front of the nose. The blood can run backwards and not out the front. The second one is a good description of a completely retracted thin eardrum due to extreme Eustachian tube dysfunction but this diver's drums were normal. The following case history is republished with permission of the Editor of the *New Zealand Medical Journal*.

A Diver's Adventure with a Doctor

Sir, In January 1987, two days after his adventure, scuba diver Neil consulted me for reassurance because of a recurrence of pain in his right ear on diving to 60 feet (20m) on two occasions. Neil had ruptured both eardrums in January 1982 in a careless swift 70 feet (23m) dive to rescue a weight-belt. In October 1982 he was referred to me, six days after his dive because of excess pain in his right ear due to difficulty, not inability, to clear his right ear. He had a haemorrhage into his eardrum. To help this problem he had his right inferior turbinate cauterised and for four years he had no trouble with the right ear until 10 days before this recent visit. In November 1983 he attended because he had ruptured his left eardrum four days previously on a 40 feet (13m) descent. As is usual for normal eardrums, it was already healed but there was a red line just posterior to the handle of the malleus, the normal site for a linear tear in a normal drum to occur with barotrauma.

In January 1987 doing a 60 feet (20m) dive he had difficulty in clearing his right ear having pain all the way down from 10 feet (3m). He knew he was clearing his ear but the pain persisted until he began to ascend when the pain went. The same experience occurred with his second dive to 60 feet (20m), so being a careful diver he consulted the local diving doctor at this popular resort. Therein lay his mistake.

He was given a thorough and alarming counselling session. The doctor told him that he had "2 bubbles of water behind the ear (drum)" and that he had "an infection in his ear." For this was prescribed Amoxil and another tablet (unknown) to dry up the water. A second visit four days later after no diving and two days prior to my consultation, he was told that the infection was still present and it was due to "the bone behind the cheekbone pressing on the Eustachian tube, or enlarged adenoids." He was told to give up his seven year duration diving as he "had a great chance of having a burst eardrum when he was on the bottom and he would drown without his lungs filling with water."

He was told that "when the eardrum bursts on the bottom you will lose your balance and you will not know which way was up or down". You need an operation before you can dive again. Neil quite rightly pointed out that if he was on the bottom he could not go down any further and that there was only one way he could go. He further volunteered that if he inflated his buoyancy compensator he could only go upwards. He was politely ignored.

"You will get water in your throat and go into a panic and drown." Neil again pointed out that he had already ruptured his eardrums on two previous occasions and had had no trouble. (In the November 1983 case records is his description of the rupture of the left eardrum and it is as follows "I was diving to 40 feet (13m) and on descent I cleared the right ear easily but not the left. It went bang and there was a hiss. I was a bit dizzy (non-directional) but sat on the bottom and it went away. I gathered my scallops, ascended and climbed into the boat without any trouble." The doctor was unimpressed and told him he needed a hearing test as he would be deaf. "The eardrum is sucked in instead of out." He reiterated his graphic story of how Neil would drown. A nice friendly doctor.

Needless to say Neil had the "fear of God" put into him and hastily returned home and rang for an appointment two days after the last visit to that doctor. Examination showed very normal eardrums; not even a reddened handle of malleus which often persists for a week or more after diving. The nose was clear. The Toynbee test of Eustachian tube function was normal and he could Valsalva with minor effort. The postnasal space was clear of adenoids which normally start atrophying about age eight years, and have gone completely by age 15 years.

However he did have some dental disorders. There were two missing molar teeth and one premolar in the lower right jaw. There was an overeruption of the corresponding teeth in the upper jaw. His six anterior lower teeth were worn flat from grinding of the teeth. Audiometry showed no change since 1983. The diagnosis was minimal nasal congestion (of which Neil admitted to having on his last dive) giving some difficulty in clearing his right ear. This occurred in an ear sensitized by the irregular dental occlusion giving preclinical mandibular dysfunction syndrome earache. The extra stretching of the eardrum plus the cold of the water gave him the earache. The resort from which he dived has an average surface sea-water temperature of 68°F (20°C) in February an 59°F (15°C) in August for the years 1959-72. Neil was reassured, counselled, and is off diving again but as a precaution has been given a nasal spray and pseudoephedrine tablets to use if he has trouble inflating his ears.

The misdiagnosis could be due to Neil's misinterpretation of what the doctor said but repeating the story while I wrote it down he did not alter it at all, apart from changing his "ruptured eardrum" to the doctor's "burst eardrum." The second reason could be the usual doctor reaction that earache equals ear infection with some congestion of the blood vessels of the eardrum due to excessive barotraumatic stretching, proving his point.

One wonders about the "two bubbles of water" and the tablets for the water in the ear. Neil does have in the right outer ear canal $^{1}/_{8}$ inch (2mm) localised rounded exostoses close to the eardrum and these might account for this statement.

The Concise Oxford Dictionary has a bubble defined as a "spherical envelope of liquid enclosing air etc" and not the reverse however. So we are left with a bewildered, seven years experienced diver thoroughly upset by a well-intentioned unknowledgeable young doctor. My experience is that less than 20% of divers who rupture their eardrums suffer from vertigo and even then it does not appear to be anywhere near the severity of that due to alternobaric vertigo.

INDEX

Abscess in ear, **16**
Adenoids, 10, **61**, 95
Adhesions, 11
Alcohol,10
Allergy, 50
Allergic rhinitis, 10, **50,** 95
Alternobaric vertigo, **44**
Aqueduct of the Cochlea, **102**
Asthma, 4
Audiogram, **38**, **40**
Auto-inflation, 20, 71, 75, 76.

Balancing apparatus, **81**, 83, 86
Bends, 36, 39, 42
Blister, **34**
Blocked ear, **91,** 93
Boil in ear, **27**
Bony ear swellings, 31
Boyles Law, **82**

Carbogen, **40**
Carbon dioxide, **35**, 66
Carbon monoxide, **35**, 66
Caribbean ear, **27**
Catarrh, middle ear, **19**
Catarrh, nasal, 10, 48
Catarrh, throat, **60**
Cautery, 13, 81, 95
Cerebro-spinal fluid, 36,37,102
Claustrophobia, **2**
Clearing ears, 20,21,**70**,71,75,76
Clenching teeth, **63,** 87
Cochlea, **35**
Cold, Common, 10, 48, **49**
Cold sores, **59**
Congestion, nasal **69**, 77, 95
Coral ear, **27**
Cotton wool bud, 17, 27, **29**, 30
Cough, **60**
Counselling, **70**
DCP Meter, **37**
Deafness
  inner ear, 36
  nerve, 35
  sensorineural, 35, 36
Decompression illness, 39, 40

Decongestant, 12, 55
Deep X-ray therapy, **69**
Dizziness, 35, **79**

Ear syringing, 26, 100
Ear toilet, **29**
Eardrops, **101**
Eardrum, 3, 14
  perforation, 15, 16, 82
  rupture of, 6, 12,23,82,104
Earplugs, 17, **18**
Emboli, 37
Emphysema, 4
Epilepsy, 4
Eustachian cushion, 6,10,**74**
Eustachian tube, 2,**5**,69,95
Exercises, jaw muscle, **89**
Exostoses, 25, **31**, 105
External otitis, **27,** 100

Fitness, 2, 4
Frenzel Manoeuvre, 72, 73
Fungal infections, 27

Glue ear, 15, 17, 19, 68

Hair cells, **35,** 101
Hay fever, **51**
Headaches, jaw, **64**
  migraine, **67**
  scuba, **65**
  sinus, **66**
Hearing Cells, **35,** 101
Hot bathing, 27, **28**

Impacted wisdom teeth, 24, 79, 88
Inferior turbinates, 69,81,95
Inflation, ear, 20,24,71,75,76
Inhalation, **56**
Inner ear, **35**, 101
Itchy ears, 29, 91

Jaw joint, **62, 87**
Jaw joint disorder, **62, 87**
Jaw muscle, **63**, 64, 89

Levator Palati, **75,** 96
Lip sores, **59**
"Locking", 11, 24, **73**, 96
Lump in throat, **60**

Mastoid, 15
Mastoid cavity, **47**
Mastoid infection, 17
Meniere's disease, **43**
Middle ear catarrh, 17, 19
Migraine, **67**
Molar teeth, **64**
Mouth piece, 88, 92
Mucous ear, 15
Mucous membrane, 10
Mylohyoid muscle, 96
Myringoplasty operation, **99**
Myringotomy, **17**

Nasal allergy, **50**
Nasal catarrh, **52**
Nasal soak, **55,** 58
Nasopharynx, 11, 73, 94
Nausea, **79,** 84
Nerve deafness, 20,35
Noises in ear, **92**
Nose bleed, **52**, 57
Nose blowing, 50
Nystagmus, 39

Occlusion, 87
Olive oil, 27
Orientation, 35, 45, 80
Otitic barotrauma, 12, 17, 23
Otitis media, **15**
Outer ear canal, 14, 26, 99
Outer ear infection, 27,100
Outer ear squeeze, **32**
Oval Window, 15,24,37,39,42
Oxygen, 39, 41

Palato-glossus muscle, 96
Palato-pharyngeus muscle, 99
Paraffin, soft **18,** 32
Patulous Eustachian tube, **95**
Penis, 10
Perilymph, 103
Pinna, 26, 100
Pneumothorax, 4
Politzerisation, 20, 21, 99
Polyps, nasal, **51**
Post nasal drip, 61
Prevention:
  abscess in ear, 17
  deafness, 20, 42
  exostoses, **32**
  lip sores, **59**
  otitis media, **17**
  outer ear infection,**30**
  reverse squeeze, **45**
  sinus, 3, **52**
  sinus squeeze, **57**
Squeeze, **24**
Swimmer's nodes, 32

Q tips, **28**
Queen Hood, 18, **19**, 30, 47

Recompression chamber, 39, 40
Referred pain, **62**
Reissner's membrane, **37**, 42
Reverse squeeze, 44
Rhinitis, 10
  allergic, 10, 11
Round Window rupture, 13, 20, 37, 43, 85, 101
Roydhouse Manoeuvre, **75**

Seasickness, **42**
Secretory otitis media, 15, 19
Sensorineural deafness, 12, 35, 36
Septoplasty, 48, 95
Septum, nasal, 10, **48**, 74
Shimmer, 46
Singapore ear, **27**
"Sinus", 3, 10, 13, 25, 48, 52
Sinus Squeeze, **57**
Sinuses, 2, 49, 51, 57
Sinusitis, 3, **54**
Skip breathing, **66**
Sneezing, 50
Soaks, nasal, **55**, 56
Soft palate, 8, 9, **75**
Sports injury, **2**
Sports Medicine, **2**

Squeeze:
middle ear, **21, 68**
outer ear, **32**
reverse, **44**
sinus, **57**
Stapes, 44, 81
Stirrup bone, 44, 81
Submariners, **6**
Sudden deafness, 36, 42
Submucous resection of Septum, **48**
Sunbathing, 10, 70
Swimmers nodes, **31**
Syringe, 26, 100
Teeth, 63, 64,88
Temporalis jaw muscle, **63**, 64
Temporo-mandibular joint, **63**, 100
Tensor Palati muscle, 75
Tensor Tympani muscle, **99**
Throat catarrh, **60**
Tinnitus, 103
Toynbee Manoeuvre, 2,20,70,95
Twitch, **76**

Uvula, **75**

Valsalva Manoeuvre, 2, 20, 37, 68, **71,** 76
Vascular headache, 67
Vaso-constrictor, 48
Vaseline, 32
Venturi effect, **56**
Vertigo, 35, 43, 79, 86
Vestibular membrane, **37**
Virus infection, **39**
Voluntary tubal opening, 71,**73**
Vomero-ethmoid spur, **10**, 95
Vomiting centre, **84**

Waterproofing, **26**
Wax, **26**, 28, 97, 99
Whiteout, **44**
Wick, **27**
Wisdom tooth, 62, 87

X-ray sinuses, **49**, 54

Yawn, **75**, 96

GLOSSARY

The three parts of the ear are the Outer Ear, Middle Ear and Inner Ear.

ABSCESS IN EAR: An infection in which pus collects inside the middle ear.
ADENOIDS: These are a form of tonsil tissue situated at the top of the throat, at the back of the nose, in the naso-pharynx, adjacent to the Eustachian tube opening.
ALTERNOBARIC: A built-up word to describe a form of vertigo or rotary dizziness brought about by changes in barometric pressure. Alterno: alternating and baric: pressure.
AUDIOMETER: A high-tech machine for scientific measuring of how well a person can hear.
BAROTRAUMA: A word from baro: meaning pressure, and trauma: meaning actual damage to the tissues of the body. Otitic being the adjective for the ear, plus Barotrauma is the technical name for the middle ear "squeeze"; Otitic Barotrauma.
CARTILAGE OR GRISTLE: The white elastic tissue coating the joint surfaces of bones and forming certain parts of the skeleton. In the Dogfish the whole of the skeleton is cartilage, hence no bones. The pinna is cartilage covered by skin.
CATARRH: Superficial inflammation caused by infection, chemical or physical irritation.
CEREBRO-SPINAL FLUID: Surrounds and circulates around the brain and spinal cord.
CLEAR THE EARS, INFLATE THE EARS, AUTO-INFLATION, AERATION OF THE MIDDLE EAR: Is forcing air up the Eustachian tube.
CONGESTION: Excessive or undue accumulation of blood inside the blood vessels, a reversible condition but causes thickening of mucous membranes.
DECONGESTANTS: Any way of reversing congestion by medicines acting on the blood vessels, decreasing their size and emptying them of blood.
DERMATITIS: A skin infection.
EARDRUM: The thin but normally strong vibrating membrane dividing the outer ear from the middle ear.
EMPHYSEMA: Chronic lung problem causing shortness of breath.
EUSTACHIAN TUBE: Is an air tube which joins the back of the nose (nasal cavity) to the middle ear.
INFLAMMATION: Reaction of the tissues to damage or infection.
MASTOID: Is the part of the skull behind and above the ear, usually honeycombed with air cells which are continuous with the middle ear.
MUCOUS MEMBRANE: Is the mucous producing lining of the nose and other passageways leading off the nose.
NASO-PHARYNX: Is that part of the throat behind the nose and above the hindmost part of the palate, or soft palate.
OTITIS: Is inflammation of the ear, either outer (external) or middle (media).
PURULENT: Is any infection or inflammation in which pus is formed.

RESPIRATORY SYSTEM: Exchanges oxygen for carbon dioxide in the lungs and also repeats this in the capillaries of the muscles and other tissues.
ROUND WINDOW and OVAL WINDOW: Are the two openings between the middle and inner ears (both normally closed with a membrane).
"SINUS": A folk medicine term give to the nasal condition of attacks of nasal blockage, excess secretions with or without face-ache and headaches.
"SQUEEZE": Is due to the increasing pressure on the body because of the height of water above the diver, the commonest effect being on the eardrum.
VASCULAR: All the blood vessels, arteries, veins and capillaries, form the vascular system.
VASOCONSTRICTOR: The act of shrinking or constricting blood vessels.
VERTIGO: Is a special form of giddiness with a rotating or spinning feeling.
WISDOM TEETH IMPACTION: Because these are the last teeth to come through the gums and because they are the last in the row they do not always grow straight towards the surface. If they are tilted they may grow into the adjacent tooth causing impaction and usually earache.If impacted they will never erupt fully if at all.

Medicine List

Acetic acid as 3% is known as Vinegar. In a pure form (Glacial Acetic Acid) is in eardrops to acidify the skin.
Acyclovir Cream for herpes labialis or cold sores on the lip.
Atomol: nasal decongestant spray.
Avomine: anti-seasickness, anti-nausea, anti-vertigo.
Betahistine Hydrochloride (Serc) normalizes metabolism in the inner ear.
Buccastem is Stemetil absorbed under the tongue. It is an anti-nausea, anti-vomiting and anti-vertigo medicine.
Calcium Carbonate: chalk is commonly used for indigestion as in Quickeeze or Soda mints.
Cerumol: An effective dissolver or softener of earwax with a turpentine base.
Chlordiazepoxide (Librium): A tranquillizer but specific for preventing grinding of teeth when asleep.
Dexamethasone: A cortico-steroid as an anti-stress hormone.
Dramamine: Anti-seasickness, anti-nausea tablet.
Drixine: nasal decongestant spray.
Eskornade: a tablet for clearing nasal congestion.
Eurax lotion: Useful for eczema and scabies but very effective against itching. Cures ear itch within 5 seconds.
Glacial acetic acid is pure acid and needs diluting for use. Present in Vosol, Spirit/Acid drops.

Iodine and Boracic Powder: Iodine is still a powerful antiseptic but needs diluting with Boracic or Boric acid powder; 60 mg Iodine to 10 grams of fine powder.

Jadit: Anti-fungal solution which can be used in the ear with precautions. Its fluid base can irritate inflammed skin.

Kenalog in Orabase (Squibb) is an ointment that sticks to wet surfaces. Contains anti-fungal, anti-bacterial and a cortico-steroid anti-stress hormone.

Locorten-vioform paste (Ciba) is an antiseptic skin rash reliever but is impermeable to all of the skin rays.

Marzine: anti-seasickness, anti-nausea, anti-vertigo.

Mycostatin dusting powder can be blown into the ear canal as an anti-fungal.

Nasal decongestant: sprays - Atomol, Drixine, Otrivine tablets- Sudafed, Sudamyl, Eskornade.

Neomycin: an antibiotic for use on skin found in many eardrops. Similar to Soframycin.

Otrivine: nasal decongestant spray.

Oxypentifylline (Trental - Hoechst): Thins the blood so that is travels quicker through the capillaries thus increasing the oxygen supply.

Paraffin comes as liquid or oil, soft as in vaseline or hard as in candlewax.

Prednisone: A cortico-steroid anti-stress hormone like dexamethasone.

Quickeeze: medicated chalk for indigestion.

RVPlus (Paul B. Elder Co): Contains titanium-coated mica impermeable to all the sun's rays.

Scopaderm: A skin patch to prevent seasickness.

Serc: Betahistine Hydrochloride used for Meniere's disease.

Soda mints: Calcium carbonate or chalk flavoured with peppermint, for indigestion.

Spirit drops to dry the ear or prevent any outer ear infection. Used to toughen the skin after the ear infection is cured. Contains 5% acetic acid to restore the skin acidity.

Sofradex: Ear drops containing antibiotic (Soframycin) and cortico-steroid (dexamethasone) for outer ear infection.

Soframycin: Antibiotic for use on skin as for Neomycin.

Sudafed: pseudo-ephedrine which clears the nose by shrinking the blood-vessels, both arteries and veins.

Sudamyl: as for Sudafed.

Titanium Dioxide: A white pigment that prevents all the sun's rays from passing through.

Trilafon: anti-seasickness, anti-nausea and anti-vertigo drug.

Vaseline: Trade name for soft paraffin and can be hardened by mixing with hard paraffin.

Vosol: Eardrops for the prevention of swimmers ear or outer ear infection. Also good for treatment of outer ear infection. Better with extra 2% acetic acid.

BIBLIOGRAPHY
Scuba Diving Scientific Publications by Noel Roydhouse

1970 The ear and water sports. Aust J Sports Medicine 1970; 3,4: 11-27.

1971 The ear and water sports, in proceedings FINA 2nd medico-scientific meeting, Carney D, Ed. Dublin 1971.

1973 Scuba Diving and the Ear, Nose and Throat. Roydhouse P. Auckland, 1973 (Book).

1974 The ear and scuba diving. Congress Proceedings, the XXth World Congress of Sports Medicine, Melbourne, 1974: 465-470.

1975 Exostoses and external otitis, letter from an ENT Surgeon. NZ Family Physician 1975;2,4: 45-46.

The quiet world that hurts the ears, Interview. Thursday Magazine, NZ 1975;Jan 16: 40-43.

Scuba divers, nasal congestion and how to "clear the ears". NZ J of Sports Medicine 1975; 3,2:13-18.

Scuba Diving and the Ear, Nose and Throat. 2nd Edition. Roydhouse T. Auckland 1975 (Book).

1976 The squeeze, the ear and prevention. NZ J of Sports Medicine 1976; 4,2: 18-25.

1977 The jaw and scuba diving. J Otolaryng Soc Australia 1977;4: 162-5.

Ear and Nose problems in Sports Medicine. ed. Williams JGP, Sperryn RN. Chpt 21, p341-351, London, Edward Arnold, 1977.

1978 The squeeze, the ear and prevention. South Pacific Underwater Medicine Soc J 1978; Jan-March: 34-38.

Ear Nose and Sinus Problems. In Basic book of Sports Medicine. Chpt 32, ed. Olympic Solidarity of the International Olympic Comm, Rome, Vatican Polygot Press 1978.

Nausea vertigo and diving. NZ J of Sports Medicine 1978; 6,4: 23-25.

Referred aural symptoms in diving. Patient Management 1978; December: 55-61.

1980 Nausea, vertigo and diving. Auditory Soc newsletter 1980;5,3:188-191.

Swimmers ears. Canadian J Appl Sport Sciences 1980;5,3:188-191.

1981 Sudden deafness and scuba divers, correspondence. NZ Med J 1981; 93:131.

Underwater Ear and Nose Care. Roydhouse N. Auckland, 1981 (Book).

1982 The differential diagnosis of a case of underwater ear pain. SPUMS J 1982; July-Sept: 29-31.

Immediate management of diving emergencies, correspondence. Brit J Sports Medicine 1982;16: 272.

1983	Vertigo in divers. correspondence. Brit J Sports Medicine 1983; 11,2:40.
	Compressed-air cylinder tap defect. South Pacific Underwater Medicine Soc J 1983;3:8-9.
	Sudden deafness in divers. South Pacific Underwater Medicine Soc J 1983;4:17.
1984	"My stupid ears", imaginary consultation. South Pacific Underwater Medicine Soc J 1984; 14, 1:9-11.
	Sudden sensorineural deafness in SCUBA divers. In current topics in Sports Medicine, Proceedings of the World Congress of Sports Medicine, Vienna 1982, 1089-1098. ed. Bachl N. Prokop L. Suckert R. Wien, Urban and Schwarzenberg, 1984.
	Emergency ascent. Correspondence. Undercurrent 1984; 9,8:6.
1985	1001 Disorders of the Ear Nose and Sinuses in Scuba divers. Canadian J Appl Sport Sciences 1985; 10, 2: 99-103.
1987	A diver's adventure with a doctor, correspondence. NZ Med J 1987; 100, 823: 297-298.
1988	Diver's ear pain or Claws 2. South Pacific Underwater Medicine Soc J 1988; 19:32-33.
	Tinnitus and diving. Auckland Tinnitus Association Newsletter 1988; 32:4-7.
	Prevention of middle ear barotrauma in scuba diving. J. Otolaryngol Soc Australia 1988; 6,2:113-115.
	Asthma and diving. NZ J Sports Medicine 1988;16:84-87.
1990	Prevention of barotrauma. Conference Proceedings of the Commonwealth and International Conference on physical education, sport, health, dance, recreation and leisure 1990; 4,2; 158-163.
1991	Tinnitus and sudden sensorineural deafness in a scuba diver. Patient Management 1991; August 75-76.
	Problems de'l'oreille, du nez et des sinus propres a la plongee sous-marine autonome. Spectrum International (Fr) 1991; 34,2: 13-16.
	Oor-, Neus - en Sinusproblemen Bij autonoom Apparaatduiken. Spectrum International (Dutch) 1991;34,2:13-16.
	Prevention of barotrauma. In: Sade, J. editor. Proceedings of a conference on the Eustachian tube and middle ear diseases, October 1990. Amsterdam, Kugler & Ghedini. 1991.

"I can make anyone's ears suitable for scuba diving" said the author at an international meeting on the Eustachian tube in Geneva, Switzerland in 1989.

"I have just told a diver to give up" said the chairman"

"Here is a copy of my book. This will help you."

Wade Doak, Tutukaka, on the original book.
"This book has been brought about by the great increase in the number of people of all ages who are taking up the recreation of underwater swimming. Noel Roydhouse is internationally known for his work in Sports Medicine having lectured on the 'Ear and Water Sports' in many countries.
This is a practical manual, well illustrated with the text explained more fully with many case histories. As well as describing in easily understood terms the troubles that can arise underwater, details of how to prevent them are given in full."